St. Louis Community College

Library

5801 Wilson Avenue
St. Louis, Missouri 63110

MODERN
DC-TO-DC
SWITCHMODE
POWER
CONVERTER
CIRCUITS

MODERN
DC-TO-DC
SWITCHMODE
POWER
CONVERTER
CIRCUITS

Rudolf P. Severns
Senior Staff Engineer
Applications Engineering
SILICONIX, INC.

· Gordon (Ed) Bloom
Senior Partner and Engineering Specialist
e/j BLOOM Associates

Van Nostrand Reinhold Electrical/Computer Science and Engineering Series

VNR VAN NOSTRAND REINHOLD COMPANY
———————————————————— New York

Library of Congress Catalog Card Number: 84-12026
ISBN: 0-442-21396-4

Manufactured in the United States of America

Published by Van Nostrand Reinhold Company Inc.
135 West 50th Street
New York, New York 10020

Van Nostrand Reinhold Company Limited
Molly Millars Lane
Wokingham, Berkshire RG11 2PY, England

Van Nostrand Reinhold
480 Latrobe Street
Melbourne, Victoria 3000, Australia

Macmillan of Canada
Division of Gage Publishing Limited
164 Commander Boulevard
Agincourt, Ontario MIS 3C7, Canada

15 14 13 12 11 10 9 8 7 6 5 4 3 2

Library of Congress Cataloging in Publication Data

Severns, Rudolf P.
    Modern DC-to-DC switchmode power converter circuits.

    (Van Nostrand Reinhold electrical/computer science and engineering series)
    Bibliography: p.
    Includes index.
    1. Electronic apparatus and appliances—Power supply.
2. Microelectronics—Power supply. I. Bloom, Gordon.
II. Title. III. Title: Modern DC-to-DC switchmode
power converter circuits. IV. Series.
TK7868.P6S47 1984   621.3815'32   84-12026
ISBN 0-442-21396-4

FOR
JOY AND DIANA
AND
THE UNSUNG HERO
OF
ELECTRONIC DESIGN,
THE POWER SUPPLY ENGINEER

Van Nostrand Reinhold
Electrical/Computer Science and Engineering Series
Sanjit Mitra–Series Editor

HANDBOOK OF ELECTRONIC DESIGN AND ANALYSIS PROCEDURES USING PROGRAMMABLE CALCULATORS, by Bruce K. Murdock

COMPILER DESIGN AND CONSTRUCTION, by Arthur B. Pyster

SINUSOIDAL ANALYSIS AND MODELING OF WEAKLY NONLINEAR CIRCUITS, by Donald D. Weiner and John F. Spina

APPLIED MULTIDIMENSIONAL SYSTEMS THEORY, by N. K. Bose

MICROWAVE SEMICONDUCTOR ENGINEERING, by Joseph F. White

INTRODUCTION TO QUARTZ CRYSTAL UNIT DESIGN, by Virgil E. Bottom

DIGITAL IMAGE PROCESSING, by William B. Green

SOFTWARE TESTING TECHNIQUES, by Boris Beizer

LIGHT TRANSMISSION OPTICS, Second edition, by Dietrich Marcuse

REAL TIME COMPUTING, edited by Duncan Mellichamp

HARDWARE AND SOFTWARE CONCEPTS IN VLSI, edited by Guy Rabbat

MODELING AND IDENTIFICATION OF DYNAMIC SYSTEMS, by N. K. Sinha and B. Kuszta

COMPUTER METHODS FOR CIRCUIT ANALYSIS AND DESIGN, by Jiri Vlach and Kishore Singhal

HANDBOOK OF SOFTWARE ENGINEERING, edited by C. R. Vick and C. V. Ramamoorthy

SWITCHED CAPACITOR CIRCUITS, by Phillip E. Allen and Edgar Sanchez-Sinencio

SOFTWARE SYSTEM TESTING AND QUALITY ASSURANCE, by Boris Beizer

MODERN DC-TO-DC SWITCHMODE POWER CONVERTER CIRCUITS, by Rudolf P. Severns and Gordon E. Bloom

Foreword

As each area of technology with a potential for significantly impacting any major segment of the electronics industry evolves, it often is accompanied by the development of a succession of new circuits. Each new circuit indeed appears different, employing different components in differing configurations, and claims an assortment of distinct features of "improved performance." Without a considerable investment of laboratory time to construct, evaluate, and compare each candidate circuit, it usually is difficult to realistically appraise the relative merits of one approach over another. It often is even more difficult to identify the underlying principles which point up basic similarities and differences. Such is the situation in the new and rapidly expanding area known as *electronic power processing* or *switching mode power supplies*.

The area of switching power supplies has been spurred by the need for power sources of higher performance, smaller volume, and lighter weight in order to achieve compatibility with the shrinking size of all forms of communication and data handling systems, and particularly with the portable battery-operated equipment in everything from home appliances and handtools to mobile communication equipment. Static dc-to-dc converters and dc-to-ac inverters provide a natural interface with the new direct energy sources such as solar cells, fuel cells, thermoelectric generators, and the like, and form the central ingredient in most uninterruptable power sources. Such solid-state power conditioners operating with internal conversion frequencies of tens to hundreds of kilohertz are emerging as the new mainstay of most power supplies for computers and communication systems, as they have already served over the last two decades for the space programs of the world.

A vast amount of circuit technology for switching power supplies has been generated in the past decade, and reported in numerous technical journals, conference proceedings and trade magazines. Our understanding of this rapidly evolving technology can be greatly enhanced by the introduction of some meaningful classification technique which assists in recognizing and focusing on fundamental elements and characteristics. In this way, the most important but often subtle aspects of a particular subject are brought to the surface so as not to become lost in the less important details. It has been found that such a classification scheme is indeed beneficial when working with the host of dc-to-dc converters which have evolved over the past two decades.

The authors have shown, for the first time, that a large class of the static dc-to-dc converters which are widely used in power processing applications fall within one of two families, namely, the buck-derived family and the boost-derived family. Each family is characterized by an equivalent averaged circuit model which exposes quite clearly the fundamental properties essential to the respective classes of converters. For example, the control-to-output transfer property of the boost-derived family of converter circuits exhibits non-minimum phase properties, i.e. a right-half S-plane zero, and two moving poles as the duty-cycle control signal is varied. The buck-derived converters, on the other hand, exhibit a zero and a double-pole which are both stationary. The differences in the fundamental properties that divide these two converter families are essential to the determination of a control strategy that assures loop stability and optimizes the dynamic responses.

Based on this classification scheme, a third class of hybrid converters that share the properties of both buck and boost converters has become evident. The familiar buck/boost converter and Ćuk converter, for example, belong to the hybrid family. Through the introduction of the principle of duality relationship and bilateral inversion, a framework is established which makes evident natural relationships between converter families and enables a methodical transfer of knowledge gained about one family of converters to its dual counterpart. As a result of this classification structure and the duality relationships revealed through it, many new converter topologies are suggested.

Further, the authors present the idea of integrated magnetics and show that various magnetic components in a converter can be integrated in a simple magnetic core structure. It is demonstrated that the integrated magnetic concept can be applied in a generalized fashion to all converter circuits. The simplified magnetic circuits may have a profound impact in future converter design.

I am very pleased to see that the authors have successfully brought together a vast amount of circuit technology in a refreshing and well organized manner. This book, with its emphasis on switching mode power supply design and topology selection, will assist experienced design engineers and help to bring novices up to speed. Congratulations, Rudy and Ed, for a job well done.

FRED C. LEE
Professor
Department of Electrical Engineering
Virginia Polytechnic Institute
and State University

Preface

From the everyday tools of science and business to the toys our children play with, no modern electronic product today goes untouched by the enormous strides made in microelectronic design technology over the past two and one-half decades. However, with each significant advance in micro-miniaturization of electronic systems has come a serious challenge to the designer of associated power processing networks to follow suit. These challenges are particularly difficult for those of us who must answer their call, for power processing techniques, as we know them today, are inherently discrete, by necessity physically larger than a user electronic system, and often wasteful in energy required to perform the processing or conditioning needed. This latter fact is especially true of those power processing schemes which utilize linear dissipative circuit approaches.

As the sizes of electronic systems began to shrink twenty years ago, it soon became very clear that bulky dissipative methods for power processing and conditioning would have to be replaced by more efficient and space-saving design techniques if overall product miniaturization was ever to be optimized. As a result, a new power electronics design specialty began to evolve, which has come to be known as the design of power conversion equipment using switch-mode power processing in conjunction with modern power semiconductors.

While the antecedents of this electronic design specialty clearly go back much earlier in time, it was the development of modern power semiconductors and the need for lightweight and efficient power processing equipment for space exploration programs that provided the initial impetus to explore and develop switch-mode conversion techniques. As is so often the case, it wasn't very long before military and, finally, commerical applications were being found for these switch-mode converter circuits (SPCs). In fact, today, in both dollar volume and unit numbers, the commercial use of SPCs has completely eclipsed their application in military and space electronic systems.

As a result of the widespread acceptance of SPCs, significant numbers of electronic engineers are now engaged in their design and application. It is not very surprising, given the ingenuity of the average electronic engineer, that literally hundreds of SPC circuit variations exist, all of which can be used for electrical power processing. In addition, much has been learned about the fundamental behavior of these circuits. Most of their basic behavior can be presented rather simply, and in a manner which is of great practical interest to circuit designers.

A basic educational problem facing an electronics engineer new to this field has been the scattering of information concerned with the design of SPCs in hundreds of publications, patents, technical papers, correspondence internal to electronic firms, private memoranda, and other scientific reporting media. As a result, only a small portion of the practical information related to this specialty is common knowledge and in general use.

Because there is such a limited pool of public knowledge in this design art, 95% of SPC designs now in service use one of six topologies, or variations thereof, to perform the desired power processing. While these conventional circuit approaches have served well and are useful for many applications, they represent only a few out of literally hundreds of possibilities. Also, these few circuits are not necessarily the best choice for many of the applications in which they are being used.

Another effect of limited knowledge is the poor transient response of many converter systems to both input line and output load changes. This is frequently due to a less than complete understanding by designers of small-signal modeling and analysis methods for SPC circuits. This problem is so widespread that many SPC system designers and their users automatically assume that poor transient response is an inherent characteristic of these circuits when, in actuality, nothing could be further from the truth. Significant improvements in the methods of modeling and small signal characterization of SPC circuits have been made in the last ten years and highly accurate and easily applied representations are now possible. The combination of good dynamic system models, multi-loop control schemes, and high frequency switching can provide SPCs with transient response characteristics that rival conventional linear dissipative circuits.

The wide acceptance of SPC technology has led to an incredible variety of applications, many of which impose severe requirements. More often than not, today's SPC designer is faced with performance requirements that can only be fulfilled by circuit designs which fully exploit state-of-the-art technology. The old "business-as-usual" approach, employing one or two SPC circuit variations for all applications, is no longer viable. Increasing the output power level quickly brings the designer face-to-face with component stress limitations in conventional SPC topologies. The seemingly endless quest for physically smaller SPC systems often forces designers to use power switching rates in the hundreds of kilohertz to reduce size, a practice generally unheard of until the late 1970s. When properly and thoughtfully applied, the use of alternate circuit topologies can be a powerful tool for circumventing the problems presented by circuit component limitations as well as by operation at extraordinarily high frequencies.

New and more energy efficient power components continue to be developed and are finding widespread acceptance by SPC designers. The metal oxide semiconductor field effect transistor (MOSFET) is a prime example of a new circuit element. While power MOSFETs can be successfully used in any conventional SPC circuit, alternative circuit topologies exist that take full advantage of the capabilities of these new semiconductors.

The motivation for this book is to gather as much of the current design knowledge concerning SPC circuits as possible in one place, and to present that knowledge in a direct and unified manner. Because the amount of available information is really quite voluminous, it is neither possible nor practical to restrict it to the contents of a single book. Therefore, in this text, we will restrict the discussion to the fundamental behavior and characteristics of SPCs and the many topological possibilities. The information presented herein is, in our opinion, that which is of the greatest utility to the designer of SPC circuits and systems. Therefore, we have deferred detailed discussion of component selection and the comprehensive design aspect of related system regulatory control to future texts in this power electronics series.

If we examine the present state of our understanding of SPC circuits, many gaps are readily discernible. In particular, a quantitative and accurate means for modeling and characterizing the large signal dynamic behavior of these circuits is still very much missing, although a concerted research and development effort is now in process to fill this void in our knowledge. It is very tempting to delay writing a book such as this one until all pieces of the SPC design puzzle have been found and put into place. No doubt, in a few years when the design art of SPCs has reached a level of considerable maturity, some excellent and highly definitive texts will be written on the subject. Unfortunately, many of us cannot afford to wait that long. So, despite its shortcomings, this text will still provide a broad introduction into the basic operation and characteristics of SPCs as well as into many of the more advanced aspects of SPC design.

This book is intended to be used both by electronics engineers just beginning to design SPC circuits and the old hands who have many years of experience in the field. Each will find much information which is useful for solving practical design problems. The text of the book is divided into two major sections. Section I (Chapters 1 through 11) presents the basic theory of SPC circuits as we understand them at this point in time. The presentation is straightforward with a minimum use of mathematics throughout, although some mathematical exercise is required in Chapters 2, 3, and 10 to support the ideas therein.

The discussions of Section I begin with the detailed exploration of two elementary SPC circuits to determine their large and small signal characteristics, and then builds progressively from these basic circuits to much more complex ones by using combinations of the basic circuits in conjunction with synthesis techniques.

The discussions of Section I assume the point of view that the desired power processing is voltage-related, that is, a voltage converter. This point of view could have just as well been current conversion-oriented and, in fact, we will see in Chapter 9 that these two points of view are duals electrically. The reason we have selected to pursue the voltage option is simply that this is by far the most common electrical conversion requirement encountered by SPC designers. The reader who has a need for a current converter should read Chapter 9 carefully, keeping in mind that the principles presented in that chapter can be applied to his requirement by the use of simple electrical duality relationships. Section I ends by discussing methods for comparing the relative merits of the multitude of SPC circuit topologies.

Section II (Chapter 12) is devoted to the examination of a variety of unique SPC circuits that fall into a special category of designs. This particular category encompasses converters whose primary inductive and transformer elements have been integrated (i.e., combined) into single physical assemblies. In many cases, the designs shown in Chapter 12, along with the methods behind the magnetic integration process, have never been shown or discussed before in any publication. Also contained in Chapter 12 is a brief review of magnetic fundamentals important to the understanding by the reader of the integration methods presented.

To further assist the reader, an extensive bibliography and lists of suggested reading materials has been included at the end of the book for reference and research purposes. We would like to acknowledge that only a portion of this book represents original contribution by the authors to the art of SPC design, with much of the information presented derived from the efforts of many others. The bibliography and reading list are attempts to acknowledge this fact. Of course, no practical reference list can ever be complete and, no doubt, many original contributions have been overlooked. To those individuals we have unintentionally omitted in this listing, we offer our sincere apologies for the oversight.

There are many individual researchers in the power electronics sciences who have distinguished themselves through their efforts to expand the design art of SPCs. In particular, we would like to acknowledge the work of Professor R. D. Middlebrook and his colleagues in the Power Electronics Group at the California Institute of Technology in Pasadena, California. Because of their dedicated research efforts, practical modeling and analysis of SPC systems is now feasible. Also, they have been instrumental in providing power electronics engineers with measurement tools and design techniques that have led to new SPC circuits and to a higher level of understanding of control techniques for SPCs than thought possible ten years ago.

A great deal of individual research into the science of power converter technology was required to write this text, including countless trips to the archives of many technical libraries and to facilities nationwide who maintain up-to-date records of SPC patents. Without the aid and assistance of Joy Bloom in the preparation of this material, including over a year's effort of manuscript editing and processing, this book would not have been completed in a timely manner. We are certainly in debt to this very special lady.

Both of us are practicing electronic design engineers and the creation of this book has been an education and often a revelation on the overwhelming possibilities of SPC circuit variations. If reading this book opens up for you the new vistas and possibilities which we have discovered in writing it, then we will feel our efforts have been well spent.

G. E. Bloom
R. P. Severns

Contents

11. COMPARATIVE TECHNIQUES FOR SPC SELECTION / 254

12. CONVERTERS WITH INTEGRATED MAGNETICS / 262

BIBLIOGRAPHY / 325

SUGGESTED READING AND OTHER INFORMATION SOURCES / 330

INDEX / 335

MODERN
DC-TO-DC
SWITCHMODE
POWER
CONVERTER
CIRCUITS

1. Introduction to Switchmode Power Converters

In a general sense, a power converter can be defined as a device which converts one form of energy into another on a continuous basis. Any storage or loss of energy within such a system while it is performing its conversion function is usually incidental to the process of energy translation. There are many types of devices which can provide such a function with varying degrees of cost, reliability, complexity and efficiency.

The mechanisms for power conversion can take many basic forms, such as those which are mechanical, electrical, or chemical in processing nature. We will be concerned here only with power converters which perform energy translation electrically and in a dynamic fashion, using the limited variety of components collectively illustrated in Fig. 1.1. How these circuit components are interconnected will be determined by the power translation desired. We restrict our choice of electrical components to include only inductors, capacitors, transformers, and switches, intentionally omitting resistors because they introduce undesirable power losses. High efficiency is usually an overriding requirement in most applications for electronic switchmode power converter (SPC) circuits, so resistive circuit elements must be avoided. This is not to say that resistances do not exist in SPCs. In their physically realizable forms, none of our chosen circuit components are ideal and all will introduce some resistive loss in the conversion process. However, every effort is made by SPC designers to minimize these parasitic resistances so that their presence and influence on SPC efficiency and conversion performance will not be significant. Only on rare occasions and for very special reasons are power consuming resistances introduced into the main power control paths of an SPC circuit. In auxiliary circuits, such as sequence, monitor, and control electronics of a total SPC system, high value resistors are commonplace, since their loss contributions are usually insignificant.

In SPC circuits, the semiconductor switches controlling the dynamic transfer of power from input to output are either fully ON or fully OFF, with very short transition times from one of these states to the other. Component voltage wave-

1

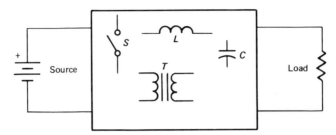

Fig. 1.1. Switchmode power converter components.

form shapes take the forms of either periodic pulse-width-modulated (PWM) rectangles, square waves, or combinations thereof. The associated current waveform shapes are usually triangular to trapezoidal. Semiconductor parts normally employed as switching elements in SPCs are fast recovery diodes, bipolar junction transistors (BJT), metal oxide semiconductor field effect transistors (MOSFET) and gate turn-off thyristors (GTO).

The most general form of the conversion function would be the transformation of a source of polyphase AC potential of one magnitude, phase, and frequency to another magnitude, phase, or frequency. With this concept, a DC input or output of an electrical power converter system can be viewed as AC with zero frequency and phase.

The discussions in this book will be limited to those circuits which perform a dynamic DC to DC (DC-DC) power conversion function. This is not nearly as restrictive a view as it might first appear, since there are a great many applications for a DC-DC SPC. There are also a large number of applications for AC to DC (AC-DC) conversion. This latter function is usually implemented by preceding a DC-DC converter circuit with a simple rectifier-filter network, a typical example of which is shown in Fig. 1.2. DC to AC and AC to AC power converter systems frequently use AC-DC and DC-DC networks as subsystem elements working together to obtain the desired output characteristics.

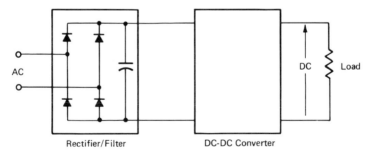

Fig. 1.2. Typical implementation of AC-to-DC conversion.

It is also interesting to note that many motor control systems use circuit components and topologies that closely resemble DC–DC converters. If we consider the uniform rotary and, in some cases, linear output from a motor to be equivalent to a DC output, then the motor and its controller network can be viewed as a DC–DC or AC–DC converter. It follows then that a great deal of the design technology related to DC–DC converters is directly applicable to motor control electronics. For reasons such as this, a study of DC–DC converters using SPC techniques is very worthwhile and has wide applicability to other related power control disciplines.

Many virtues are expected of electrical power converters. They must be highly efficient, occupy little volume, be low in cost and weight, and have a long service life with no failures. Often their outputs must be regulated to close tolerances, even though their inputs may vary over a very wide range of values. The family of SPC circuits, which we will explore in this text, are able to fulfill these requirements remarkably well, provided their designs are well executed.

Depending on application needs, a DC–DC SPC circuit could be very simple in form, using only one SPDT switch, one inductor, and one capacitor. On the other hand, it may be very complex, utilizing many components to accomplish the required DC conversion function. Regardless of the complexity of a particular SPC circuit, the following key idea always applies:

Key Idea #1

THE INDUCTORS AND CAPACITORS OF A DC–DC SPC MUST, IN SOME MANNER, FORM A LOW-PASS FILTER NETWORK.

The justification for this idea is the need for a DC output voltage with a minimum of superimposed AC ripple from the switching action of an SPC. Since the internal nodal or branch waveforms of an SPC are at some point pulsating with both AC and DC components, a low-pass filter is required to prevent the undesirable AC components produced by SPC switching action from appearing at the output terminals of the converter.

It is not always easy to isolate the low-pass inductive and capacitive filter elements of an SPC from a casual inspection of its circuit diagram, since the effective filter inductor element (or elements) is often separated from the filter capacitor element (or elements) by switches and/or transformers. In the modeling processes to be discussed in Chapters 2, 3, and 10, we will invariably derive an equivalent low-pass filter network for each converter type. The most common circuits have a single-section low-pass filter. However, there are some special SPC circuits (such as the Ćuk SPC) which contain multiple-section equivalent filter networks.

Two of the simplest SPCs are shown in Figs. 1.3 and 1.4, along with physical realizations for their SPDT switch functions (Figs. 1.3B and 1.4B). The circuit

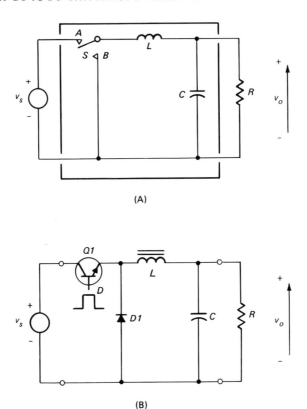

(A)

(B)

Fig. 1.3. The buck converter.

in Fig. 1.3 is called a buck (step-down) converter because the output voltage is always less than the input voltage. The circuit of Fig. 1.4 is referred to as a boost (step-up) converter because its output voltage is always greater than the input voltage.

One point should be made here. The output capacitor (C) in the buck converter is not absolutely essential for the circuit to function properly, since its inductor (L) and load resistor (R) form a low-pass filter by themselves and, in theory at least, L can be made arbitrarily large in value to produce a small ripple current in R. However, even the presence of a small amount of output capacitance will greatly reduce the size of L required to obtain a DC output voltage across R with a minimal AC ripple. In actual practice, C is rarely omitted. For this reason, we will assume that the basic buck converter will usually possess an output filter capacitor. On the other hand, the output capacitor of the boost converter cannot be omitted, since the output current will always be pulsating even if the inductor is infinitely large. Thus, to obtain a low-ripple output voltage,

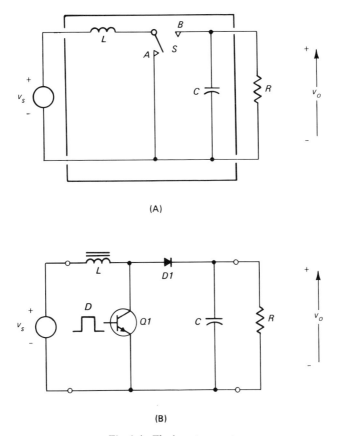

(A)

(B)

Fig. 1.4. The boost converter.

a filter capacitor is always necessary across the output terminals of a boost converter.

We will now put forward a number of additional key ideas regarding SPCs in general.

Key Idea #2

ALL KNOWN DC-DC SPC CIRCUITS CAN BE DERIVED FROM COMBINATIONS OF BUCK AND/OR BOOST CONVERTERS ALONG WITH SOME FORM OF TRANSFORMATION FUNCTION.

The transformation function is shown in a general fashion in Fig. 1.5A, and can be characterized as an ideal transformer element, as illustrated in Fig. 1.5B. Because of its ideal properties, such an element can transform DC as well as AC.

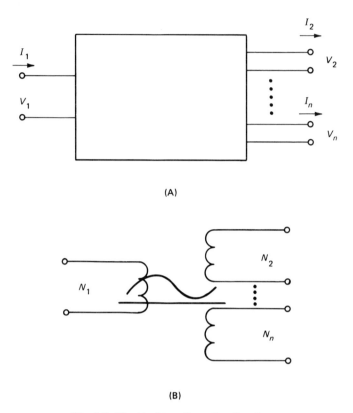

(A)

(B)

Fig. 1.5. The ideal transformation function.

Obviously, an ideal transformer element would be very difficult to realize in actual practice. Fortunately, perfection is not usually required in transforming voltages and currents within SPCs, and there are many practical design approximations which serve quite well. This subject of both DC and AC transformation will be treated in more detail in Chapter 4.

The justification for the assertion in Key Idea #2 is strictly empirical, i.e., no SPC circuit known at this point in time is so complex that it cannot be reduced to a combination of these elementary circuits. In the discussion that follows, the truth of this key idea is assumed and the structure of this book is organized in accordance with this premise. Chapters 2 through 4 will examine the two basic SPC converter circuits and transformation elements in great detail. That material will then be used in later chapters to examine more complex converter structures.

A third key idea is directly related to the second one:

Key Idea #3

THE PROPERTIES OF BASIC CONSTITUENT CONVERTERS (i.e., buck and/or boost) ARE RETAINED IN ANY COMPLEX DERIVA-TIVE CONVERTER SYSTEM.

This key idea holds for the majority of complex SPCs, with the exception of a few converter circuits which use tapped inductors and transformers with equal turns ratios between windings. In some of these special converters, basic behavior patterns are modified to some extent. This subject is discussed in Chapter 10. Except for these special exceptions, the above idea is sound and the general properties of a complex SPC can be readily predicted from a knowledge of its basic constituents. This ability to directly relate circuit properties of many SPCs is a valuable tool in comparing the merits of more complex circuits.

For example, if one needs an SPC with inrush and short-circuit current control capabilities like those exhibited by a buck SPC, he must use an SPC topology that is buck-derived. By its very nature, an unmodified boost-derived converter cannot provide such current-control features. When one attempts to modify a boost-derived converter to include these features, one of two things will happen. The modified circuit will either become a buck-derived SPC or it will exhibit either buck-derived or boost-derived characteristics, depending on input voltage and output current magnitudes. Examples of SPCs which can have this type of dual personality are discussed in more detail in Chapter 9.

A fourth key idea which is of practical importance to an SPC designer is:

Key Idea #4

NO SINGLE SPC CIRCUIT TOPOLOGY IS ELECTRICALLY IDEAL FOR ALL APPLICATIONS. IT IS THE PARTICULAR APPLICATION WITH ITS INDIVIDUAL REQUIREMENTS THAT DETERMINES WHICH CIRCUIT WILL BE THE BEST TO USE.

This may seem to be a very obvious, almost trivial statement. However, even a brief look at the claims (and often the practice) within the power electronics industry will reveal that this simple truth is not universally appreciated nor accepted. If this key idea were not true, then this book would be dedicated to the discussion of the one most perfect circuit for all situations.

SPC circuits do exist which have been very carefully synthesized to have as many desirable and as few undesirable characteristics as possible. The Ćuk SPC circuit and its variations discussed in Chapter 7 are good examples. To an amazing degree, concerted efforts such as these have succeeded in producing designs which are near optimum for many applications. Despite these successes, however, SPC circuits such as the Ćuk versions are still not optimum for *all* applications.

What makes Key Idea #4 so important is the fact that even relatively small changes in SPC circuit topology often lead to significant alterations in component stresses, or changes in dynamic control-related characteristics, such that desired conversion properties might be severely compromised. On the other hand, SPC circuit topology alteration can be an effective tool for circumventing design difficulties created by the limitations and nonideal properties of components. Therefore, it is up to the designer to make the proper choice of alteration, always keeping in mind the pros and cons of his actions.

SPCs may be categorized by their general output characteristics. Sometimes, this can be most helpful to the user in defining the application requirements that must be met by an SPC. Fig. 1.6A shows a general converter system in block form, along with output voltage (V) and current (I) defined as illustrated. Fig 1.6B gives the possible directional combinations of V and I, where each quadrant

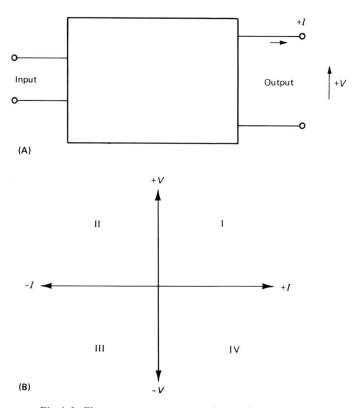

Fig. 1.6. The converter output operating quadrant concept.

is identified by sign. In a simple DC–DC converter with unidirectional power flow (input-to-output), the output is constrained to lie in either quadrant I or quadrant III of Fig. 1.6B, depending on the desired output voltage and current polarity. This is an example of single-quadrant operation. In battery charger/discharger applications, the polarity of the converters' output potentials are fixed in sign, but their output currents may be either positive or negative in directional flow. Thus, their output characteristics must lie either in quadrants I and II or quadrants III and IV of Fig. 1.6B. This is an example of two-quadrant operation, where the power flow must be bidirectional. AC output of an SPC would also be a two-quadrant operational system. In the most general case, a converter with AC output and bidirectional power flow capability must operate in all four quadrants of Fig. 1.6B.

Before proceeding to the detailed circuit discussions of the next chapters, let us pause and review the terminology that we will be using in the text. Unfortunately, even among power electronics engineers, there is no universally accepted terminology for SPC circuit names and their operational modes, much less their origins. The present authors have taken the liberty of adopting terminology for this text which pleases us, and admittedly some of our choices are arbitrary. If the reader has adopted other conventions of terminology, he should have little difficulty in interpreting the descriptions to follow.

SPCs are often referred to as switching regulators, especially by their users. Most SPC *systems* do contain auxiliary control or regulating electronics to provide a constant output voltage with varying output load and input voltage—hence, the appellation *switching regulator*. In this text, we will discuss the power stages of switching regulators and refer to them as SPCs.

The word *converter* is used here in its general sense, referring to a physical system that is capable of converting electrical power of one form to another form, regardless of whether its inputs and/or outputs are AC or DC in nature. This same term is sometimes used in a more restricted sense to indicate that its input and output characteristics are the same form of power, i.e., AC–AC or DC–DC. The name *inverter* is used when DC–AC or AC–AC power conversion is implied. Since we are restricting the discussions of this text to DC–DC SPCs, these latter distinctions will not be needed.

Alternative names exist for the *buck* SPC. It is not uncommon to hear this basic SPC circuit referred to as a *voltage step-down converter*, *buck regulator*, or a *chopper converter*. The term *chopper* comes from the dynamic action of the SPDT switch within the converter which produces a "chopped" voltage at the input to its low-pass filter network.

Boost converters and their basic first-order derivatives also have other names by which they are known. These names include *ringing choke*, *flyback* or *voltage step-up* converters. The term *flyback* is particularly confusing, since the same term is often applied indiscriminately by designers in discussing three very different SPC circuits—the basic boost converter structure, a converter system which combines a buck and a boost converter in cascade, and a transformer-coupled version of a cascaded buck and boost converter network. In this text, the name *flyback* will be applied only to the last of the three alternative SPC circuits.

Buck and boost SPCs are frequently referred to as buck and boost regulators, respectively, since they are primarily used as the power stages of such regulation systems. For simplicity, we have chosen not to include these names as terms of the discussions of this text as was stated earlier in this chapter.

The adjectives *buck-derived* and *boost-derived* will be used quite frequently in the chapters to follow. When applied to a given circuit under consideration, they imply that it has been derived in some relatively direct manner from either a basic buck or a basic boost converter, respectively, and that the general properties of the parent converter have been retained. These descriptive terms will become very important as the process of evolution of large numbers of different converter topologies begins and the need to categorize them becomes evident in order to simplify the processes for application selection.

Lastly, the *switching frequency* (f_s) of an SPC circuit will be defined to mean the rate at which all of its circuit switches complete one full switching cycle of their operation. It is important that the switching frequency of an SPC not be confused with AC output voltage or *current ripple frequencies* (f_r) for, more often than not, f_r is an integer multiple of f_s. In the two basic SPC circuits to be examined in the next two chapters, their input, as well as output, ripple current frequencies are equal to the basic SPC switching frequency. However, in many other SPCs, the ripple frequencies of voltages and currents can be many times higher than the associated SPC switching frequency.

2. The Buck Converter

The first of the elementary switchmode power converters (SPCs) that we will examine is the buck converter. The goal of the analyses of this chapter, as well as the next one, is to develop mathematical and equivalent circuit models that adequately represent circuit operation and to quantitatively determine essential electrical characteristics required for proper design. In the analysis process, we will derive most of the expressions from which numerical values for circuit performance and component stresses may be computed.

A highly desirable feature of any electrical circuit model is that it be readily manipulated using the techniques of linear circuit analysis and synthesis familiar to electrical engineers. Unfortunately, SPC circuits are nonlinear and discontinuous by their very nature, and, as such, are very cumbersome to analyze directly using standard linear circuit theory. A major part of this chapter is therefore devoted to "linearizing" the buck converter circuit model so that it can be easily analyzed. The linearizing process does involve a slight compromise in accuracy, but for most of the quantities we are seeking, the accuracy of the model is more than adequate for normal design purposes.

The models to be derived must possess many other virtues as well. An SPC is rarely used as a power processing system unto itself. There are usually additional networks such as sequence, monitor, and control electronics along with input and output filters for EMI noise suppression. Sometimes an SPC will be used in conjunction with other converter circuits. For reasons such as these, any SPC models of practical use must be able to accommodate any additional circuit elements as easily as possible.

Another constraint placed on our SPC models is the need to recognize the non-ideal character of real circuit components, such as inductor winding resistance and core loss or the equivalent series resistance (ESR) of filter capacitors, and to be able to insert these parasitic elements into the models without undue complexity in the incorporation process.

Finally, our models must be sufficiently easy to use, so that they can be applied easily by the SPC designer in his work. A model, regardless of how accurate it may be, cannot be a general purpose design tool if many hours of calculation are required to generate the desired information. An SPC is expected to be efficient in its translation of power, and no less can be expected of its modeling abilities from a time viewpoint. The models that we will derive in this chapter and in Chapter 3 meet these requirements amazingly well.

The following modeling discussion tries to minimize the use of mathematics. As a result, some basic assumptions have been glossed over and most theoretical derivations have been omitted. For the reader who may be interested in these mathematical details, an excellent exposition of state-space averaging is given in Refs. 1 and 2 in the text bibliography of this book.

We begin by taking a very simple look in Section 2.1 at the circuit operation and waveforms for an ideal buck converter. Then, in Section 2.2, equations for the large-signal steady-state behavior for both conduction modes of this SPC will be presented. Section 2.3 will review the circuit's large-signal dynamic character-ization. In Section 2.4, we will derive the small-signal characterization as well as a circuit model for the continuous mode of the buck SPC, using the state-space averaging technique. Parasitic elements will be included in this small-signal model. Finally, in Section 2.5, we will examine the operation of an ideal buck converter in a discontinuous mode situation and, once again using the methods of state-space averaging, derive the describing equations and a circuit model for this second operating mode.

The discussion of Section 2.4 is intended to demonstrate the procedure of state-space averaging. While this procedure is somewhat involved (and is really an overkill for the simple case of an ideal buck SPC), it is important to familiarize ourselves with these methods, as they are the only practical means for modeling both the discontinuous mode of a buck SPC and all modes of operation of a boost converter.

A basic assumption throughout the analysis to follow is that the output voltage ripple of the SPC is small. This is equivalent to saying that the output filter cor-ner frequency (f_c) is much lower than the switching frequency (f_s). This as-sumption is one of the key means by which a linear SPC circuit model can be formulated and is fundamental to the principles of state-space averaging. Fortu-nately, the assertion that f_s is much greater than f_c is valid for any SPC producing a DC output with low AC ripple content.

2.1. BUCK CONVERTER OPERATION

An idealized circuit for the switch states of a buck SPC is shown in Figs. 2.1B and 2.1C, along with a practical realization for the SPDT switch in Fig. 2.1D. Corresponding typical waveforms for the ideal buck converter of Fig. 2.1A are diagrammed in Figs. 2.2 and 2.3.

During the switching cycle, the SPDT switch (S) moves between positions A and B, so that the circuit topology alternates states between that of Fig. 2.1B and that of 2.1C. At the beginning of a switching cycle (t_0), S is in position A, the

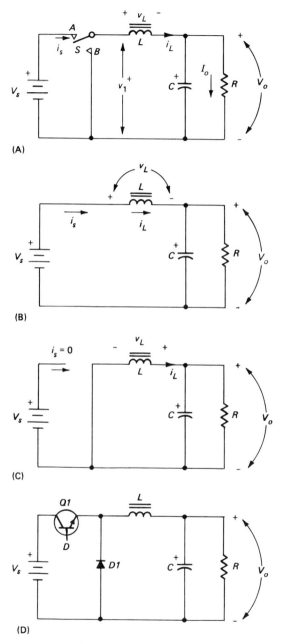

Fig. 2.1 The buck converter.

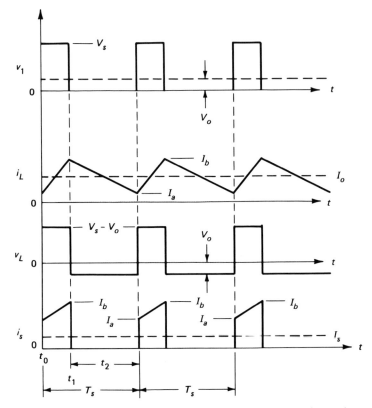

Fig. 2.2. Buck converter waveforms for the continuous mode of operation.

input voltage (V_s) is greater than the output voltage (V_o), and the current (i_L) in the inductor (L) ramps upward during the interval t_1, as shown in Fig. 2.2.

The inductor current will continue to ramp upward until S changes to position B. When S moves to position B, at $t = t_1$, $V_s = 0$ and the voltage across L changes sign with i_L ramping downward. Two inductor current possibilities now exist at the end of interval t_2; either $i_L = 0$ or $i_L = I_a$, some nonzero positive value. In other words, either all of the energy stored in L is delivered to the load between t_1 and t_2 or some energy remains as S moves back to position A.

This defines the two basic operating conditions for the buck SPC and leads to a key idea about SPC operating modes:

MODE 1. Continuous inductor current during the entire switching cycle (continuous mode operation).

MODE 2. Discontinuous inductor current during the switching cycle (discontinuous mode operation).

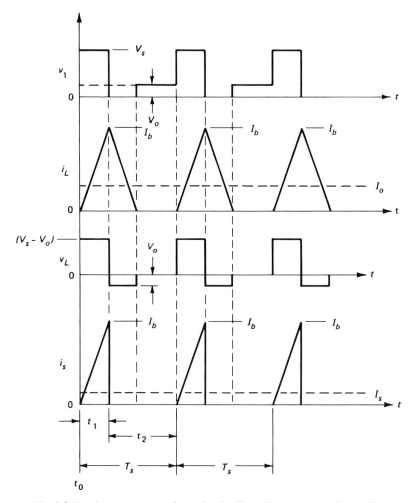

Fig. 2.3. Buck converter waveforms for the discontinuous mode of operation.

Key Idea #5

THE CONDUCTION MODE OF THE INDUCTOR CURRENT IS A FUNDAMENTAL FACTOR IN DETERMINING THE ELECTRICAL CHARACTERISTICS OF ANY SPC.

If one were to place two topologically identical SPCs in separate black boxes, the only difference between them being that the value of the inductor of one is chosen so the converter operates in the continuous mode and the other inductor chosen such that its SPC operates in the discontinuous mode, a series of tests on each SPC to determine output electrical characteristics would conclude that the two boxes contained entirely different converter circuits. The inductor current conduction modes of the basic SPCs (as well as any complex derivatives) have a profound effect on their conversion characteristics and are as important to know as their topological structures when determining circuit performance.

The *duty cycle* (D) is defined as:

$$D \equiv \frac{t_1}{T_s} = t_1 f_s . \tag{2.1}$$

For fixed values of L and D, the mode in which the buck converter operates is determined by the value of R. As R is increased in ohmic value, the operating mode will change from continuous to discontinuous. On the other hand, if R and D are fixed, the operating mode will change from continuous to discontinuous as the inductance of L is decreased. As f_s is increased, the minimum value of L to maintain the SPC in a continuous mode of operation will decrease in direct proportion to the change in f_s.

As this discussion proceeds, it will become apparent that, for many applications such as those where high-power or multiple outputs from an SPC are required, maintaining the SPC in a continuous mode of operation is very desirable, even for large variations in load resistances. To avoid using an excessively large value of L to maintain continuous current levels at light output loads, a "swinging" choke is sometimes employed. A swinging choke is an inductance whose value is a function of the current passing through it, being large at low current and getting progressively smaller as the current is increased. While this technique can be very useful, a swinging inductance will complicate the analysis of the SPC by making f_c of its associated filter variable with load resistance. This may make the stabilization of any associated regulatory control loop somewhat more difficult.

One can deduce many of the properties characteristic of a buck SPC and its more complex derivatives by examining its electrical waveforms, such as those plotted in Figs. 2.2 and 2.3.

V_o is the average voltage value of v_1 so this converter can be simply viewed as a voltage chopper followed by a low-pass filter. Voltage v_1 is composed of both AC and DC components. Since its DC voltage component is the desired output of the converter, the low-pass filter must remove any AC component of this time-varying potential. This implies that f_c of the output low-pass filter must be much less than f_s. Because V_o is the average of v_1, V_o must *always* be less than the input DC voltage (V_s) to this SPC.

Notice also that the waveform for v_1 differs in shape for each of the two modes of operation. In the continuous mode, $v_1 = V_s$ during t_1, and $v_1 = 0$ during the entire period t_2. The average value of $v_1(V_o)$ is determined by D only and, ideally, V_o would be independent of R in this mode. In the discontinuous mode, however, v_1 can be either zero or equal to V_o during time t_2. The length of time spent by v_1 in a zero-value condition during t_2 is determined by R. The value of V_o is now dependent on R as well as D. A close examination of the v_1 waveforms in Figs. 2.2 and 2.3 shows that, for the same value of V_s and D, V_o was increased in the discontinuous mode. Figs, 2.2 and 2.3 have been scaled so that both SPC output power levels are the same in value.

The input current (i_s) is always pulsating, regardless of the operating mode of the buck SPC. Most applications limit (often severely) the amount of pulsating current that can be drawn by an SPC from its source of power. In normal practice, some form of a low-pass filter network is inserted between the source of power and the input terminals of the SPC. As will be shown in Chapter 10, the presence of such a filter must be accounted for at an early stage of SPC design and modeling process. Failure to do so can result in a very unpleasant surprise in the form of self-oscillations when the SPC and the input low-pass filter circuit are united.

The average value of i_s is determined both by V_o and R. In an ideal buck converter with no power losses, efficiency (η) of power conversion is 100%. Therefore,

$$V_s I_s = \frac{V_o^2}{R}, \qquad \eta = 100\%. \tag{2.2}$$

For the normal case with a constant value of V_o, $V_o^2 = K$.
In this instance,

$$I_s = \frac{K}{V_s R}. \tag{2.3}$$

Here, the average input current value is inversely proportional to both V_s and R. However, the pulsating value for i_s is determined by i_L which, in turn, is a function of R, but *not* V_s. As V_s is varied and D altered to maintain a constant V_o, the width of i_s is altered, but *not* its maximum amplitude I_b.

The current i_s is the input switch current, and it can be seen from Figs. 2.2 and 2.3 that, for a constant value of input power, the peak magnitude of i_s is much higher when the buck SPC is operating in a discontinuous mode of operation. In a practical realization of the SPDT switch as shown in Fig. 2.1D, both the BJT (*S1*) and the diode (*D1*) must possess a much higher peak-power handling capability if SPC operation in the discontinuous mode is desired. For this and many other important reasons, continuous mode of operation of buck SPCs is normally preferred in high-power applications, even though their inductances must be larger and therefore, heavier and certainly more expensive.

The output filter cutoff frequency is defined as

$$f_c = \frac{1}{2\pi\sqrt{LC}} \qquad (2.4)$$

and it would appear that L can be chosen to be just large enough to maintain the SPC in a continuous mode, with C then selected to provide the degree of output filtering desired. Unfortunately, capacitors are not perfect electrical elements and some power will be dissipated in their ESR. The ESR voltage drop produces an output ripple voltage that can only be reduced by lowering ESR magnitude or the dynamic capacitor current causing the ripple. The size of C is more often determined by the ripple current magnitude than by the value desired for f_c. Making L larger reduces the output ripple current value which, in turn, reduces C. In actual design practice, there is usually a tradeoff made in selecting L and C for weight and size plus cost reasons. In high-power converters, such tradeoffs normally result in a value selection for L which places the SPC circuit operating well into the continuous mode at full load.

Ideally, the output current (I_o) of the buck SPC is not a pulsating quantity since the ripple content of i_L is small in the continuous mode. This is a convenient feature for many applications and, frequently, additional output filtering can be either dispensed with or made quite small.

2.2 STEADY-STATE LARGE-SIGNAL CHARACTERIZATION

To design a converter, it is necessary to quantify many of the circuit characteristics, such as the AC and DC currents flowing through its circuit elements, the voltages across these elements, the input-to-output voltage transfer ratio as a function of D, and many other factors.

The following discussion presents equations and graphs that describe the large-signal steady-state (DC) characteristics of a buck SPC. The analysis presumes an ideal buck SPC, i.e., no parasitic elements such as inductor series resistance (R_L) or capacitor ESR (R_c) are included. Despite this limitation, the equations presented still have more than adequate accuracy for practical design purposes. For an SPC to be efficient, R_L and R_c must be made small. Designers often go to great lengths to minimize lossy parasitic elements when engineering SPC circuits. Therefore, their omission from the following model derivation is justified as a matter of good design practice.

Figure 2.4 illustrates the buck SPC for which the equations in Tables 2.1 and 2.2 apply. An input filter network has been included ($L1$, $C1$), and it is presumed that $L1$ possesses sufficient inductance so that I_s is essentially DC. On the output side, $C2$ is assumed to be high in capacitance, so that the AC component of I_L flows only through this filter element. Both of these assumptions are consistent with practical application requirements of SPCs for both low input current ripple and small output voltage ripple.

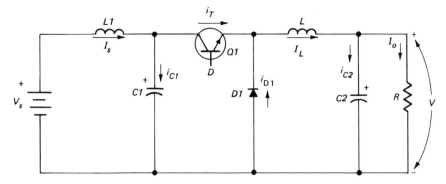

Fig. 2.4. Definition of the buck converter component currents.

For the equations shown in Tables 2.1 and 2.2, and later in Table 2.3, the following definitions apply:

$$D \equiv D_1 = \frac{t_1}{T_s} \quad \text{(switch conduction duty cycle)} \qquad (2.5)$$

$$M = \frac{V}{V_s} \quad \text{(input-to-output DC transfer ratio)} \qquad (2.6)$$

$$\tau_L \equiv \frac{L}{RT_s} \quad \text{(normalized inductor time constant)} \qquad (2.7)$$

$$I_o = \frac{V}{R} \quad \text{(DC output current)} \qquad (2.8)$$

TABLE 2.1A. Definitions of Variables for the Buck Converter Equations.

τ_L	L/RT_s
T_s	Switching period
τ_{LC}	L_c/RT_s
L_c	Critical inductance
τ_{LC}	$(1 - M)/2$
D_1	Switch ON duty cycle
D_2	Diode conduction duty cycle
M	V/V_s
I_a	Minimum inductor current
I_b	Maximum inductor current

TABLE 2.1B. Buck Converter Equations for Continuous Inductor Current.

Dependent Variable	$f(M, \tau_L, V, R)$	$f(D_1, \tau_L, V, R)$
M	M	D_1
D_1	M	D_1
D_2	$1 - M$	$1 - D_1$
$I_a\,(I_{L\,(\text{min})})$	$\dfrac{V}{R}\left[1 - \left(\dfrac{1-M}{2\tau_L}\right)\right]$	$\dfrac{V}{R}\left[1 - \left(\dfrac{1-D_1}{2\tau_L}\right)\right]$
$I_b\,(I_{L\,(\text{max})})$	$\dfrac{V}{R}\left[1 + \left(\dfrac{1-M}{2\tau_L}\right)\right]$	$\dfrac{V}{R}\left[1 + \left(\dfrac{1-D_1}{2\tau_L}\right)\right]$
$I_{L\,(\text{avg})}$	$\dfrac{V}{R}$	$\dfrac{V}{R}$
$I_{L\,(\text{rms})}$	$\dfrac{V}{R}\left[1 + \dfrac{1}{12}\left(\dfrac{1-M}{\tau_L}\right)^2\right]^{1/2}$	$\dfrac{V}{R}\left[1 + \dfrac{1}{12}\left(\dfrac{1-D_1}{\tau_L}\right)^2\right]^{1/2}$
$I_{C1\,(\text{rms})}$	$\dfrac{V}{R}\left\{M\left[(1-M) + \dfrac{1}{12}\left(\dfrac{1-M}{\tau_L}\right)^2\right]\right\}^{1/2}$	$\dfrac{V}{R}\left\{D_1\left[(1-D_1) + \dfrac{1}{12}\left(\dfrac{1-D_1}{\tau_L}\right)^2\right]\right\}^{1/2}$
$I_{C2\,(\text{rms})}$	$\dfrac{V}{R}\left(\dfrac{1-M}{\sqrt{12}\,\tau_L}\right)$	$\dfrac{V}{R}\left(\dfrac{1-D_1}{\sqrt{12}\,\tau_L}\right)$
$I_{T\,(\text{avg})}$	$\dfrac{VM}{R}$	$\dfrac{VD_1}{R}$
$I_{T\,(\text{rms})}$	$\dfrac{V}{R}\left\{M\left[1 + \dfrac{1}{12}\left(\dfrac{1-M}{\tau_L}\right)^2\right]\right\}^{1/2}$	$\dfrac{V}{R}\left\{D_1\left[1 + \dfrac{1}{12}\left(\dfrac{1-D_1}{\tau_L}\right)^2\right]\right\}^{1/2}$
$I_{D1\,(\text{avg})}$	$\dfrac{V}{R}(1-M)$	$\dfrac{V}{R}(1-D_1)$
$I_{D1\,(\text{rms})}$	$\dfrac{V}{R}\left\{(1-M)\left[1 + \dfrac{1}{12}\left(\dfrac{1-M}{\tau_L}\right)^2\right]\right\}^{1/2}$	$\dfrac{V}{R}\left\{(1-D_1)\left[1 + \dfrac{1}{12}\left(\dfrac{1-D_1}{\tau_L}\right)^2\right]\right\}^{1/2}$
$I_{s\,(\text{avg})}$	$\dfrac{VM}{R}$	$\dfrac{VD_1}{R}$

$$D_2 \equiv \frac{t_2}{T_s} \quad \text{(diode conduction duty cycle).} \tag{2.9}$$

The equations in these tables have been stated in two different ways—first as a function of D_1 for closed-loop calculations, and second as a function of M for open-loop operation. For the continuous mode, $D_1 = M$, so that the two columns

TABLE 2.2A. Buck Converter Equations for Discontinuous Inductor Current.

Dependent Variable	$f(M, \tau_L, V, R)$	$f(D_1, \tau_L, V, R)$
M	M	$\dfrac{2}{1+\sqrt{1+8\tau_L/D_1^2}}$
D_1	$M\sqrt{\dfrac{2\tau_L}{1-M}}$	D_1
D_2	$\sqrt{2\tau_L(1-M)}$	$\left(\dfrac{4\tau_L}{D_1}\right)\left(\dfrac{1}{1+\sqrt{1+8\tau_L/D_1^2}}\right)$
I_a	0	0
I_b	$\dfrac{V}{R}\sqrt{\dfrac{2(1-M)}{\tau_L}}$	$\left(\dfrac{V}{R}\right)\left(\dfrac{4}{D_1}\right)\left(\dfrac{1}{1+\sqrt{1+8\tau_L/D_1^2}}\right)$
$I_{L(avg)}$	$\dfrac{V}{R}$	$\dfrac{V}{R}$
$I_{L(rms)}$	$\dfrac{V}{R}\left[\left(\dfrac{8}{9}\right)\left(\dfrac{1-M}{\tau_L}\right)\right]^{1/4}$	$I_b\left[\dfrac{D_1+D_2}{3}\right]^{1/2}$
$I_{C1(rms)}$	$\dfrac{V}{R}\left[M\sqrt{\dfrac{8}{9}\left(\dfrac{1-M}{\tau_L}\right)}-M^2\right]^{1/2}$	$I_b\sqrt{\dfrac{D_1}{3}-\dfrac{D_1^2}{4}}$
$I_{C2(rms)}$	$\dfrac{V}{R}\left[\sqrt{\dfrac{8}{9}\left(\dfrac{1-M}{\tau_L}\right)}-1\right]^{1/2}$	$I_b\left[\dfrac{D_1+D_2}{3}-\dfrac{(D_1+D_2)^2}{4}\right]^{1/2}$
$I_{T(avg)}$	$\dfrac{VM}{R}$	$\dfrac{VD_1}{R}$
$I_{T(rms)}$	$\dfrac{V}{R}\left[\dfrac{8}{9}M^2\left(\dfrac{1-M}{\tau_L}\right)\right]^{1/4}$	$I_b\sqrt{\dfrac{D_1}{3}}$
$I_{D1(avg)}$	$\dfrac{V}{R}\sqrt{2\tau_L(1-M)}$	$\dfrac{V}{R}\left(\dfrac{4\tau_L}{D_1^2}\right)\left(\dfrac{1}{1+\sqrt{1+8\tau_L/D_1^2}}\right)^2$
$I_{D1(rms)}$	$\dfrac{V}{R}\left[\dfrac{8(1-M)^3}{9\tau_L}\right]^{1/4}$	$\left(\dfrac{V}{R}\right)\left(\dfrac{1}{\tau_L\sqrt{3}}\right)\left[\left(\dfrac{4\tau_L}{D_1}\right)\dfrac{1}{1+\sqrt{1+8\tau_L/D_1^2}}\right]^{3/2}$
$I_{s(avg)}$	$\dfrac{VM}{R}$	$\dfrac{VD_1}{R}$

of equations result in identical values. In the discontinuous mode, however, the closed and open-loop equations are quite different. Notice also that, between the continuous and discontinuous modes, the tables show equations that are distinctly different.

In order to select the correct equation set for use, it is necessary to know the

TABLE 2.2B. Equations Normalized to τ_{LC}.

$$\frac{I_a}{I_o} = 1 - \frac{\tau_{LC}}{\tau_L}$$

$$\frac{I_b}{I_o} = 1 + \frac{\tau_{LC}}{\tau_L}$$

$$\frac{I_{L(\text{rms})}}{I_o} = \left[1 + \frac{1}{3}\left(\frac{\tau_{LC}}{\tau_L}\right)^2\right]^{1/2}$$

$$\frac{I_{C2(\text{rms})}}{I_o} = \frac{1}{\sqrt{3}}\left(\frac{\tau_{LC}}{\tau_L}\right)$$

$\Bigg\}$ Continuous Mode

$$\frac{I_b}{I_o} = 2\left(\frac{\tau_{LC}}{\tau_L}\right)^{1/2}$$

$$\frac{I_{L(\text{rms})}}{I_o} = \left[\frac{16}{9}\left(\frac{\tau_{LC}}{\tau_L}\right)\right]^{1/4}$$

$$\frac{I_{C2(\text{rms})}}{I_o} = \left[\frac{4}{3}\left(\frac{\tau_{LC}}{\tau_L}\right) - 1\right]^{1/2}$$

$\Bigg\}$ Discontinuous Mode

mode in which the converter is operating. As was pointed out earlier, for a given value of M, the conduction mode is determined by L as well as R. For this reason, we have chosen to use the normalized time constant τ_L in the equations since it is a function of the ratio of L to R. As τ_L becomes smaller, the SPC's conduction mode will eventually change from continuous to discontinuous. The critical boundary value of τ_L (τ_{LC}) for the transition between modes is:

$$\tau_{LC} = \frac{1 - M}{2}. \tag{2.10}$$

If $\tau_L > \tau_{LC}$, then the continuous mode will prevail. On the other hand, if $\tau_L < \tau_{LC}$, then the converter operates in the discontinuous mode. This relationship of Eq. (2.10) is shown in graphic form in Fig. 2.5 for ease of determination of boundaries for various values of τ_L. Keep in mind that as R and V_s are varied, a mode change in SPC operation may occur. It is always prudent to evaluate the mode boundaries over the full range of R and V_s for a particular application if single mode operation is desired.

While Tables 2.1 and 2.2 are very useful for calculating specific numerical values,

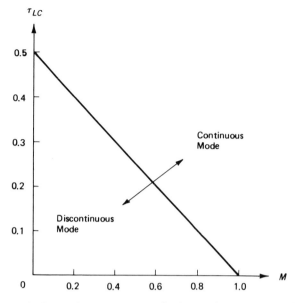

Fig. 2.5. Inductor current conduction mode boundaries.

greater visibility of buck SPC operation can be obtained by graphing the equations. For example, the relationship between M and D_1 is of great interest, and a graph of this relationship is given in Fig. 2.6.

Here we see that the DC control-to-output transfer characteristics are very different between the two modes of operation. In the continuous mode, M is a linear, monotonically increasing function of D_1, and V is not a function of circuit elements L, R, or C. In an ideal converter, V would be independent of load changes and, as such, would be self-regulating from a loading standpoint. A practical SPC converter, however, does not have perfect self-load regulation characteristics because of the presence of R_L. But if this resistance is small (the practical case), then it is reasonable to expect that a practical buck converter will display a high degree of output self-regulation for load variations, when operating in the continuous mode.

Conversely, when the buck SPC is operating in the discontinuous mode, its output is anything *but* self-regulating, with output voltage a function of D_1, R, L, and V_s. As the converter moves deeper into the discontinuous mode of operation, the control range can become quite small.

The control function DC gain (G_1) is:

$$G_1 = \frac{\partial V}{\partial D_1}. \tag{2.11}$$

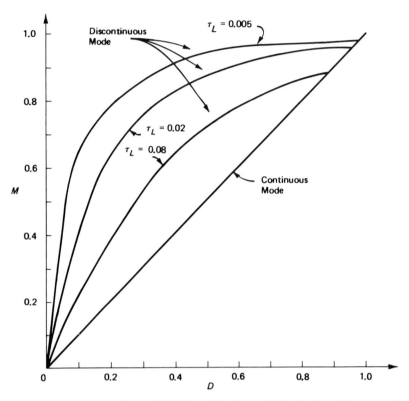

Fig. 2.6. Buck converter voltage gain (M) as a function of D for both continuous and discontinuous modes.

For the continuous mode:

$$G_1 = V_s. \tag{2.12}$$

However, in the discontinuous mode:

$$G_1 = V_s \left\{ \frac{16 \tau_L}{(\sqrt{D_1^2 + 8\tau_L})(D_1 + \sqrt{D_1^2 + 8\tau_L})^2} \right\}. \tag{2.13}$$

From Eq. (2.12) we see that G_1 is a simple linear function of V_s. But in the discontinuous mode G_1 is a much more complex function and dependent on all the circuit quantities as shown in Eq. (2.13).

Of particular interest to a designer are the rms, average, and peak currents which occur within the branches of an SPC circuit. Figs. 2.7–2.13 are graphic

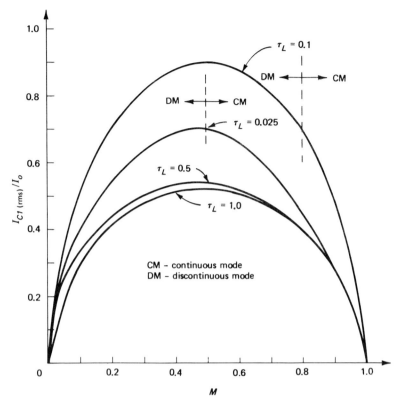

Fig. 2.7. A graph of $I_{C1(rms)}/I_O$ versus M for the buck converter.

representations of various current relationships for the buck SPC, all of which can be easily used as tools in component selection processes.

For example, to select the correct value for the input filter capacitor $C1$ of Fig. 2.4, it is necessary to know the maximum rms current that this capacitor will experience. Looking at Fig. 2.7, we see that $I_{C1(rms)}$ varies widely with the value of M. If the range of M includes 0.5, then the maximum $I_{C1(rms)}$ will occur at or near 0.5. If the range of M does not include 0.5, then by inspection of Fig. 2.7, we can see whether to use the maximum or the minimum values of M to calculate $I_{C1(rms)}$. Besides M, the value of L will effect $I_{C1(rms)}$. From Fig. 2.8 we see that if L is large compared to L_c then variations in L will have little effect on the rms current in $C1$. But, as we reduce L, $I_{C1(rms)}$ increases, slowly at first, and then quite rapidly as the discontinuous mode is entered.

Variations in output capacitor current of the buck SPC with M and L_c/L are shown in Figs. 2.9 and 2.10, respectively. For $C2$, the maximum value of current always occurs at the minimum value of M and we again see the rapid increase

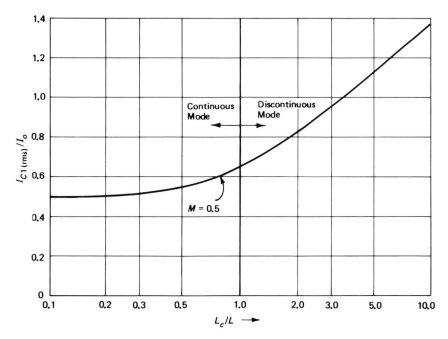

Fig. 2.8. Input capacitor rms ripple current as a function of inductor size for the buck converter.

in current as L is reduced. Figs. 2.11, 2.12, and 2.13 provide additional graphical examples of the variation of the circuit currents within an ideal buck circuit.

It is instructive to compare buck circuit component currents when L is altered to produce a conduction mode change. Such a comparison is given in Table 2.3 for $L = 2L_c$ (continuous mode) and $L = 0.5\ L_c$ (discontinuous mode). When the value of L is reduced, one would expect L to become smaller in proportion. However, $I_{L\,(rms)}$ is increasing rapidly and, since the copper losses in the inductor windings are proportional to wire turns (N) with L proportional to N^2, the copper loss increases. The larger value of AC current produced by a smaller L will also increase the core losses within the inductor. Even though the value of L has been reduced by a factor of four, the peak energy stored $(I_b^2 L/2)$ has dropped only by 13%. The net result is that the physical size of L, assuming equal losses, does not drop nearly as rapidly as its inductance value. Meanwhile, the power dissipation in $C1$ has more than doubled in the discontinuous case and the power dissipation in $C2$ has risen nearly 20 times! To handle such losses, both $C1$ and $C2$ must be made larger and their increase in size can easily offset any volume gain as a result of reduction in the size of L. For such reasons, as buck SPC output power levels go beyond 75-100 watts, discontinuous modes of operation are rarely satisfactory. At low power levels, however, circuit component sizes are

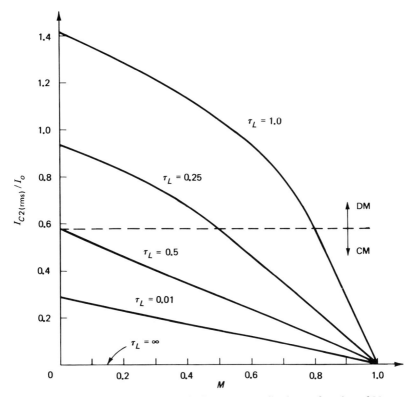

Fig. 2.9. Output capacitor rms ripple current amplitude as a function of M.

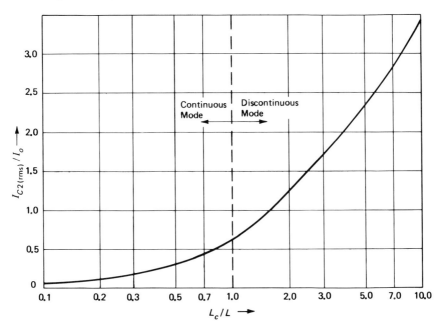

Fig. 2.10. Output capacitor rms ripple current amplitude as a function of inductance value.

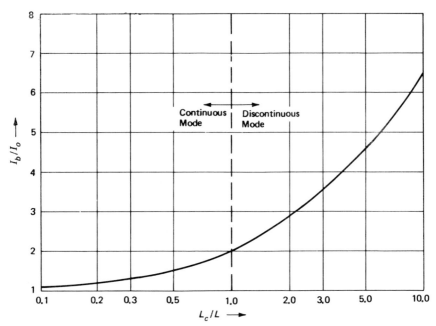

Fig. 2.11. Inductor peak current amplitude as a function of inductance value.

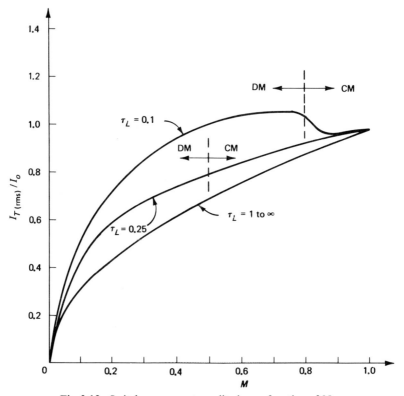

Fig. 2.12. Switch rms current amplitude as a function of M.

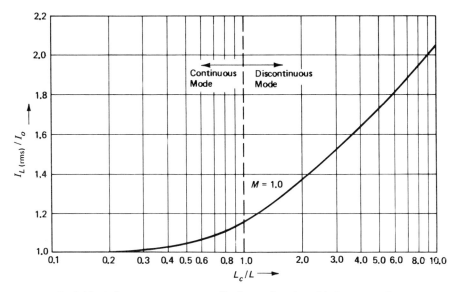

Fig. 2.13. Inductor rms current amplitude as a function of inductance value.

TABLE 2.3. Effect of Changing L_C/L.

Current	$L = 2L_c$	$L = 0.5L_c$	Ratio	Ratio^2
I_b	1.50	2.83	1.87	3.49
$I_{L(\text{rms})}$	1.04	1.37	1.32	1.73
$I_{C1(\text{rms})}$	0.54	0.83	1.54	2.36
$I_{C2(\text{rms})}$	0.29	1.29	4.45	19.8

not so directly related to their current-handling capabilities. Here the discontinuous mode is frequently used, especially where large output load ranges are possible, to avoid conduction mode changes and corresponding effects on small-signal response functions important for regulation stability.

2.3. LARGE-SIGNAL DYNAMIC CHARACTERISTICS

A major concern of the designer is how an SPC circuit will respond to step-changes in input voltage and output load current, especially when such variations are produced by open circuit or overload situations.

If input/output variations are small, then excellent analytical tools are available to the SPC designer to predict corresponding circuit responses. These small-signal models will be examined in detail in later sections of this book. However, if these variations are large, currently available analytical means for predicting the SPC circuit response are quite limited and are far from accurate. Large-signal modeling and analysis of SPCs are problems that are now beginning to receive a good deal of research attention by power electronics scientists. Unfortunately, at this writing, we will have to restrict our discussions to some fairly general observations, unsatisfactory though that may be.

If the load of a buck SPC is removed, then $\tau_L = 0$ and the output voltage V will rise toward a value of V_s. V can be limited by providing auxiliary protection elements to sense the output voltage rise and to turn off the switching elements. Following such a sequence, V will fall to zero with a time constant dependent on the values of $C2$ and any leakage resistance inherent in the circuit.

When the output of the buck SPC is short-circuited, I_L will begin to rise with its rate limited by V_s, f_s, and L. Therefore, it will normally take several switching cycles of the SPC before I_L becomes excessive. If a current sensor is used in series with L, there is usually sufficient time for auxiliary electronics to turn off the switch before I_L can produce circuit damage.

If V_s is applied rapidly to the circuit of Fig. 2.4 and $Q1$ is initially OFF, then the inrush current into the SPC is limited by the value of $C1$. By increasing the duty cycle of $Q1$ slowly, V can increase gradually and the inrush current at SPC turn-on limited to that required to charge $C1$ of the input filter. This is referred to as "soft-starting" and is an auxiliary control feature included in most SPC applications.

Conversely, when the SPC is turned off, the duty cycle of $Q1$ can be slowly reduced by auxiliary electronics to produce a gradual reduction of the output voltage. However, there is a limit to the degree of shutdown control possible using such means. The turn-off time constant of the control circuit must be longer than the time constant formed by R and $C2$. If this condition is not satisfied, the effective load-plus-filter network time constant will determine the turn-off characteristics of the system.

One of the advantages of the buck SPC and its more complex derivative converters is the ability to easily control output voltages and currents during turn-on and turn-off and under output fault conditions. As we shall see in the next chapter, the boost converter does not possess such advantageous properties.

2.4 STATE-SPACE AVERAGE MODEL FOR THE CONTINUOUS MODE

One of the most important advances of the past ten years in characterizing SPCs has been the development of accurate small-signal models for them, and a variety of modeling methods are now available. We have chosen to present only the state-space averaging technique, since it is very general, relatively easy to comprehend, and straightforward in its application in solving practical SPC modeling problems.

State-space averaging provides the SPC designer with a choice of either a mathematical or equivalent circuit model for his use. The mathematical model is very useful, since it permits the designer to calculate the values for SPC circuit voltages, currents, small-signal transfer functions, and much more. However, the mathematical model for an SPC does not readily expose the electrical processes occurring in the circuit. A deeper and more rapid understanding of the circuit functions can be gleaned from examination of an equivalent circuit model. In general, both types of models are usually needed to solve practical design problems. Two independent routes are available to the designer for developing each model. In the first, the circuit equations are manipulated until a satisfactory mathematical model is developed, and from that model, one proceeds to synthesize an equivalent circuit model. Alternatively, we could start with an equivalent circuit model, manipulate it into a satisfactory circuit form, and then generate related circuit equations. The two approaches just described are equivalent and the choice of which route to follow is really a matter of practical convenience. In the discussions to follow, we have elected to take the first route since it will be relatively easy for the reader to follow if this is his first exposure to state-space modeling. Again, for the researcher who is interested in a more complete exposition of the technique, the work by R. D. Middlebrook and his colleagues [1, 2] is highly recommended.

The procedure will follow these steps:

STEP 1. Draw the linear equivalent circuit for each state of the converter. In the case of the continuous mode, there are two states, each corresponding to one switch position.

STEP 2. Write the circuit equations for each equivalent circuit model of Step 1 in a state-variable format.

STEP 3. Average each set of equations by using the duty cycle of the switch as a weighting factor and then combine the two sets of state equations into a single set by summation.

STEP 4. Perturb the averaged equation set of Step 3 to produce DC and small-signal terms and eliminate any nonlinear cross products.

STEP 5. Transform the small-signal or AC terms from Step 4 into complex frequency domain (the so-called s domain).

At this point we will have a mathematical model and the final step of the procedure is:

STEP 6. Draw the equivalent circuit model corresponding to the mathematical equations formed by the earlier steps.

As an example of the above procedure, we will now go through the step-by-step process for the case of a simple buck SPC operating in the continuous mode.

The circuit to be studied is shown in Fig. 2.14A. When the switch (S) is in position A, the circuit assumes the topology shown in Fig. 2.14B, and when S is in position B, the network assumes the topology illustrated in Fig. 2.14C.

We continue the process by writing the equations in state-variable form for each of these two networks. The form of the equation is:

$$\dot{x} = A_{11}x + A_{21}y + b_1 v_s \qquad (2.14)$$

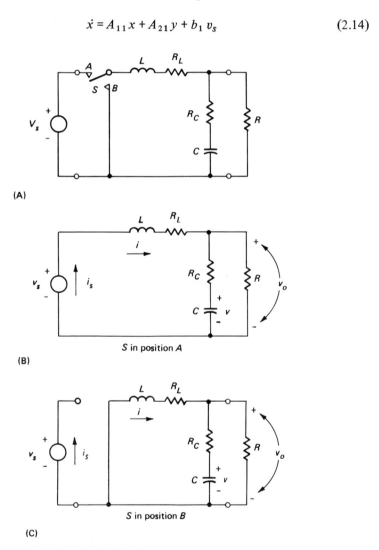

Fig. 2.14. Equivalent circuits for the buck converter in the continuous mode.

$$\dot{y} = A_{21}x + A_{22}y + b_2 v_s \tag{2.15}$$

$$z = C_1 x + C_2 y. \tag{2.16}$$

Because of their dynamic behavior, it is normal to choose inductor currents and capacitor voltages as the state variables in the equations. Alternately, inductor flux (ϕ) and the capacitor charge (q) can also be used as state variables. For the network in Fig. 2.14B:

$$\frac{di}{dt} = -\frac{1}{L}\left(\frac{R R_L + R R_C + R_C R_L}{R + R_C}\right) i - \frac{1}{L}\left(\frac{R}{R + R_C}\right) v + \frac{v_s}{L} \tag{2.17}$$

$$\frac{dv}{dt} = \frac{1}{C}\left(\frac{R}{R + R_C}\right) i - \frac{1}{C}\left(\frac{1}{R + R_C}\right) v \tag{2.18}$$

$$v_o = (R\|R_C) i + \left(\frac{R}{R + R_C}\right) v \tag{2.19}$$

$$i_s = i \tag{2.20}$$

and for the network in Fig. 2.14C:

$$\frac{di}{dt} = -\frac{1}{L}\left(\frac{R R_L + R R_C + R_C R_L}{R + R_C}\right) i - \frac{1}{L}\left(\frac{R}{R + R_C}\right) v \tag{2.21}$$

$$\frac{dv}{dt} = \frac{1}{C}\left(\frac{R}{R + R_C}\right) i - \frac{1}{C}\left(\frac{1}{R + R_C}\right) v \tag{2.22}$$

$$v_o = (R\|R_C) i + \left(\frac{R}{R + R_C}\right) v \tag{2.23}$$

$$i_s = 0. \tag{2.24}$$

Notice that we have included expressions for the input and output quantities, i_s and v_o, as these quantities will be needed to form the final equivalent circuit model.

Each of the above sets of equations is valid for a particular switch position during an operating cycle. These equations are exact and, along with boundary conditions and element values, completely describe the corresponding network of Fig. 2.14 during each portion of its switching cycle. Unfortunately, it is a very cumbersome process to derive the SPC circuit parameters directly from these two sets of equations. In addition, one rarely needs exact answers for every parameter of these state equations. To combine the two networks into a single network which can be easily analyzed using linear circuit theory, we need to *average* the equations over one cycle. The weighting factor for averaging will be the proportion of time spent by the switch S (Fig. 2.14A) in each of its two states.

For switch position A, $\Delta t = t_1$, and for position B, $\Delta t = t_2$. The weighting factors for averaging then become:

$$\frac{t_1}{T_s} = d \tag{2.25}$$

$$\frac{t_2}{T_s} = \frac{T_s - t_1}{T_s} = 1 - d = d'. \tag{2.26}$$

By averaging the state equations, we will lose some information about the SPC, but the results will be more than adequate in accuracy, especially if the output voltage ripple of the SPC is minute. Since low output ripple is normally a desirable characteristic of a DC-DC SPC, this restriction is of little analytical consequence.

The state-space averaging of a particular state variable will take the following form:

$$\langle \dot{x} \rangle = (A_{11}x + A_{12}y + b_1 v_s)d + (A'_{11}x + A'_{12}y + b'_1 v_s)d' \tag{2.27}$$

where the primed quantities represent the circuit constants associated with the state in Fig. 2.14C.

If we now apply this averaging process to Eqs. (2.17) through (2.24) the following set of equations for the state-space averaged network is obtained:

$$\frac{di}{dt} = - \left(\frac{R_L + R \| R_C}{L} \right) i - \frac{1}{L} \left(\frac{R}{R + R_C} \right) v + \frac{v_s d}{L} \tag{2.28}$$

$$\frac{dv}{dt} = \frac{1}{C} \left(\frac{R}{R + R_C} \right) i - \frac{1}{C} \left(\frac{1}{R + R_C} \right) v \tag{2.29}$$

$$v_o = (R \| R_C) i + \left(\frac{R}{R + R_C} \right) v \tag{2.30}$$

$$i_s = di. \tag{2.31}$$

We have now arrived at a single set of equations and a single model that represents the average characteristics of both states. However, the equations are not yet linear, since in Eq. (2.28) we see the term $v_s d$ and in Eq. (2.31) di, which are the products of two variables. As a practical matter, it is usually desirable to have separate solutions for AC and DC characteristics. In this equation set, both AC and DC characteristics are lumped into a single variable.

The standard technique for solving these problems is to assume that the AC terms can be made arbitrarily small compared to the DC terms. We do this by making the following assumptions:

$$v_s = V_s + \hat{v}_s \tag{2.32}$$

$$v = V + \hat{v} \tag{2.33}$$

$$i = I + \hat{i} \tag{2.34}$$

$$d = D + \hat{d} \tag{2.35}$$

$$v_o = V_o + \hat{v}_o \tag{2.36}$$

$$i_s = I_s + \hat{i}_s \tag{2.37}$$

where V_s, V, I, I_s, D, and V_o are the DC values for the circuit variables and $\hat{v}_s$, $\hat{v}$, $\hat{i}$, $\hat{i}_s$, $\hat{d}$, and $\hat{v}_o$ are the corresponding small-signal or perturbed values for the variables.

The small-signal assumption implies that:

$$\frac{\hat{v}_s}{V_s} \ll 1 \tag{2.38}$$

$$\frac{\hat{v}}{V} \ll 1 \tag{2.39}$$

$$\frac{\hat{i}}{I} \ll 1 \tag{2.40}$$

$$\frac{\hat{d}}{D} \ll 1 \tag{2.41}$$

$$\frac{\hat{v}_o}{V_o} \ll 1 \tag{2.42}$$

$$\frac{\hat{i}_s}{I_s} \ll 1. \tag{2.43}$$

Since all small-signal quantities are minute, any cross-multiplied terms, such as $\hat{v}\,\hat{d}$, are even smaller and may be ignored. If we incorporate these assumptions into Eqs. (2.28) through (2.31) and separate the small-signal and DC portions of each, the following set of equations results:

DC

$$0 = [-R_L + (R\|R_C)]\, I - \left(\frac{R}{R+R_C}\right) V + V_s D \qquad (2.44)$$

$$0 = \left(\frac{R}{R+R_C}\right) I - \frac{V}{R+R_C} \qquad (2.45)$$

$$V_o = (R\|R_C) I + \left(\frac{R}{R+R_C}\right) V \qquad (2.46)$$

$$I_s = DI \qquad (2.47)$$

AC

$$\frac{d\hat{i}}{dt} = -\left(\frac{R_L + R\|R_C}{L}\right)\hat{i} - \frac{1}{L}\left(\frac{R}{R+R_C}\right)\hat{v} + \left(\frac{D}{L}\right)\hat{v}_s + \left(\frac{V_s}{L}\right)\hat{d} \qquad (2.48)$$

$$\frac{d\hat{v}}{dt} = \frac{1}{C}\left(\frac{R}{R+R_C}\right)\hat{i} - \frac{1}{L}\left(\frac{1}{R+R_C}\right)\hat{v} \qquad (2.49)$$

$$\hat{v}_o = (R\|R_C)\,\hat{i} + \left(\frac{R}{R+R_C}\right)\hat{v} \qquad (2.50)$$

$$\hat{i}_s = D\hat{i} + I\hat{d}. \qquad (2.51)$$

The small-signal relationships of Eqs. (2.48) through (2.51) can be more easily manipulated if they are transformed into the complex frequency domain using Laplace transformation methods. These four equations will now become:

$$V_s\hat{d}(s) + D\hat{v}_s(s) = (sI + R_L + [R\|R_C])\,\hat{i}(s) + \left[\frac{R}{R+R_C}\right]\hat{v}(s) \qquad (2.52)$$

$$0 = \left[\frac{R}{R+R_C}\right]\hat{i}(s) - \left[sC + \frac{1}{R+R_C}\right]\hat{v}(s) \qquad (2.53)$$

$$\hat{v}_o(s) = (R\|R_C)\,\hat{i}(s) + \left[\frac{R}{R+R_C}\right]\hat{v}(s) \qquad (2.54)$$

$$i_s(s) = D\hat{i}(s) + I\hat{d}(s). \qquad (2.55)$$

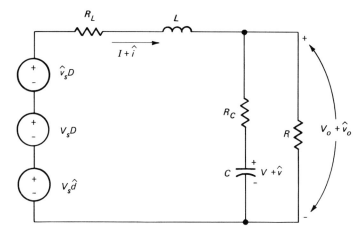

Fig. 2.15. A state-space averaged model for the buck converter.

The electrical network represented by these equations is illustrated in Fig. 2.15. For the time being, we will not elaborate on the methods of its construction. In Chapter 10, we will illustrate a general procedure for deriving the averaged circuit model of an SPC from its mathematical equations. While the circuit model in Fig. 2.15 is both simple and linear in nature, it still leaves much to be desired as a practical small-signal model. For example, the model contains generators whose values are functions of more than one variable, making it diffi-cult to embed the model within a total power processing system, which normally includes regulation control electronics as well as input noise reduction filter networks.

The next step in practical model development is to separate the DC control function from the source voltage value. One way to accomplish this is shown in Fig. 2.16A, where a variable ideal turns-ratio transformer element ($T1$) has been introduced. This transformer element has some marvelous properties, since its frequency response extends down to zero so as to pass AC and DC signals with equal facility. We indicate this transformation capability by the combination of a sinusoidal and a straight line through the transformer symbol. Also, the primary-to-secondary turns ratio of this transformer model is variable and equal to $D + \hat{d}$.

Obviously, an ideal transformer such as this one is not physically realizable. Nevertheless, it is a very useful modeling concept for our purposes.

The small-signal dependency of the variable turns-ratio of the transformer element in Fig. 2.16A can be removed if desired, and for our modeling purposes it is highly desirable to do so, as many SPC circuits have small-signal models where the generators associated with d have a complex variable dependency.

If the turns-ratio dependency of $T1$ on $\hat{d}$ is to be removed, then the model must include generator elements that account for the $I\hat{d}$ and $V_s\hat{d}$ changes in the

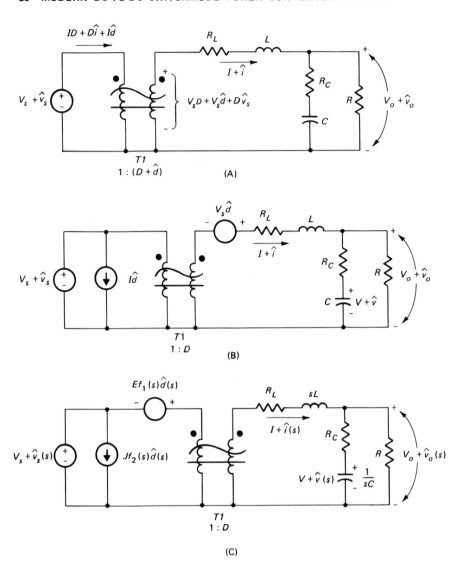

Fig. 2.16. Evolution of the final state-space averaged model for the buck converter.

primary and secondary, respectively, of the transformation element. This can be done, as shown in Fig. 2.16B, by placing appropriate voltage and current generators in the primary and secondary of $T1$. We can further simplify the model of Fig. 2.16B by translating the $V_s \hat{d}$ voltage generator to the primary side of the transformer. The resulting SPC model is illustrated in Fig. 2.16C, where:

$$Ef_1 = \frac{V_o}{D^2} = \frac{V_s}{D} \tag{2.56}$$

$$Jf_2 = \frac{V_o}{R} = I. \tag{2.57}$$

The model of Fig. 2.16C has the properties we have been seeking. All of the variables (V_s, $\hat{v}_s$, D, $\hat{d}$, etc.) have separate control elements and can be individually varied. The source and the load are clearly separated from the SPC model so as to facilitate the addition of an input filter or additional output filter circuits. The structure of the model permits its inclusion into a total power processing system with regulation controls that can alter D as a function of changes in SPC input voltage or output load. We will examine all these possibilities in more detail in Chapter 10.

The model we have evolved is somewhat arbitrary in that the primitive models we began with could have been manipulated in a different manner with the final result exposing a completely different model topology. In any case, the chosen circuit model must still conform to the constraints of Eqs. (2.44) through (2.47) and Eqs. (2.52) through (2.55), so that any model must be electrically equivalent to all others. Thus, one may choose between them as a matter of analytical convenience. The model of Fig. 2.16C meets all of our modeling needs very nicely and is now beginning to be accepted as a standard model for examining the effects of small-signal changes in control as well as input/output conditions.

Now that we have both a circuit model and a mathematical model for a buck SPC, we can begin to examine some of the circuit relationships important to small-signal operation.

For a start, we solve Eqs. (2.44) through (2.47) to obtain expressions for V_o as a function of V_s:

$$\frac{V_o}{V_s} = \underbrace{D}_{\text{ideal}} \cdot \underbrace{\left[\frac{R}{R + R_L} \right]}_{\substack{\text{correction} \\ \text{factor}}}. \tag{2.58}$$

We see that V_o/V_s, when the parasitic resistance of the inductor is included, is simply the product of the ideal voltage transfer function of the buck SPC and a correction factor that accounts for the series parasitic resistance.

This correction factor is not normally very large if the buck converter is reasonably power-efficient, and is always less than one. From the small-signal equations, one can solve for the open-loop input-to-output transfer function or the control-to-output transfer function as shown below.

$$\frac{\hat{v}_o(s)}{\hat{d}(s)} = \left(\frac{V_o}{D}\right)\left[\frac{1 + sR_CC}{1 + s\left(R_CC + [R\|R_L]\,C + \dfrac{L}{R + R_L}\right) + s^2LC\left(\dfrac{R + R_C}{R + R_L}\right)}\right] \quad (2.59)$$

$$\frac{\hat{v}_o(s)}{\hat{v}(s)} = \left[\frac{DR}{R + R_L}\right]\left[\frac{1 + sR_CC}{1 + \left[R_CC + (R\|R_L)\,C + \dfrac{L}{(R + R_L)}\right]s + LC\left[\dfrac{R + R_C}{R + R_L}\right]s^2}\right]$$

$$(2.60)$$

Figure 2.17 is a logarithmic graph of the magnitude and phase angle of Eq. (2.59) using asymptotic approximations. As shown, the equation represents a conventional two-pole low-pass filter with a zero introduced by the presence of output capacitance ESR. The magnitude of Eq. (2.59) is obviously dependent on the value of SPC duty cycle D, which will change the position of the magnitude plot, but *not* its shape.

The open-circuit input impedance (Z_{io}) of the SPC is:

$$Z_{io} = \frac{\hat{v}_s(s)}{\hat{i}_s(s)}\bigg|_{\hat{d}(s)=0} . \qquad (2.61)$$

To determine Z_{io} we could manipulate Eqs. (2.52) through (2.55) with $\hat{d} = 0$, but it is simpler to use the circuit model for the determination. Looking at Fig. 2.16C, we see that Z_{io} is the input impedance of the output filter network (including R) reflected through $T1$. From this observation we can write:

$$Z_{io} = \left[\frac{R + R_L}{D^2}\right] \cdot \left[\frac{1 + s\left[\dfrac{L}{R + R_L} + (R_C + (R\|R_L))\,C\right] + s^2LC\left[\dfrac{R + R_C}{R + R_L}\right]}{1 + s(R + R_C)\,C}\right] .$$

$$(2.62)$$

The open-loop output impedance (Z_{oo}) is defined as:

$$Z_{oo} = \frac{\hat{v}_o(s)}{\hat{i}_o(s)}\bigg|_{\hat{d}(s) = \hat{v}_s(s) = 0} . \qquad (2.63)$$

However, our analysis has implicitly assumed that $\hat{i}_o(s) = 0$. We can alter our model to cope with this situation by adding an external generator (i_g) in parallel with R, as shown in Fig. 2.18 A. Then:

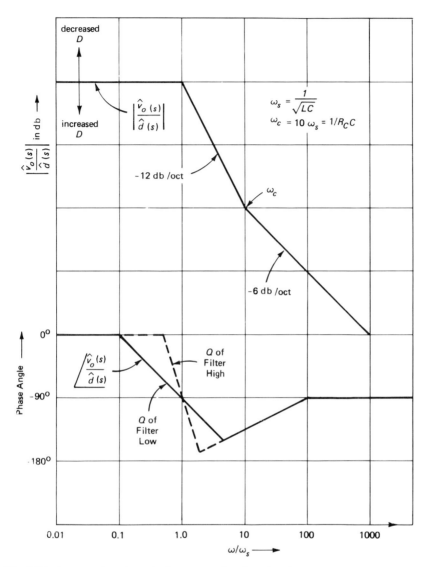

Fig. 2.17. Asymptotic gain and phase approximations for the control-to-output transfer function of a typical buck converter.

$$Z_{oo} = \frac{\hat{v}_o(s)}{\hat{i}_g(s)} \Bigg|_{\hat{d}(s) = \hat{v}_s(s) = 0} \qquad (2.64)$$

Given that $\hat{d} = \hat{v}_s = 0$, one can now redraw our circuit model to look like that shown in Fig. 2.18B. Since we are interested only in small-signal AC terms, the

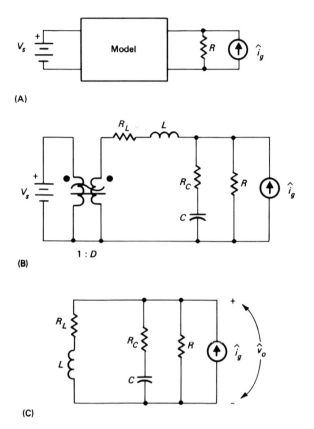

Fig. 2.18. Methods for determining Z_{oo}.

model can be further simplified as shown in Fig. 2.18C. From this latter network, we can readily determine Z_{oo} to be:

$$Z_{oo} = (R \| R_C) \left\{ \frac{(s + R_L/L)(s + 1/R_C C)}{s^2 + \left[\dfrac{R_L + (R \| R_C)}{L} + \dfrac{1}{(R + R_C) C} \right] s + \left[\dfrac{R + R_L}{R + R_C} \right] \dfrac{1}{LC}} \right\}.$$

(2.65)

An alternative method for finding Z_{oo} would be to add $\hat{i}_g$ to our original circuit of Fig. 2.14A and then derive our averaged models with its effect included; i_g could then be extracted directly from the mathematical descriptions of the resulting averaged models.

2.5. STATE-SPACE AVERAGED MODEL FOR THE DISCONTINUOUS MODE

The definition of the discontinuous mode requires that the inductor current of the buck SPC go to zero during each switching cycle. As a result, the discontinuous circuit model must assume three different states during each switching cycle in contrast to only two states for the continuous mode. In the analysis to follow, we will use an ideal buck converter circuit for simplicity.

The three networks (A, B, and C) corresponding to the three different states are shown in Fig. 2.19. By examination of each state, we can write the following state-variable network equations:

$$A \begin{cases} \dfrac{di}{dt} = -\dfrac{v}{L} + \dfrac{v_s}{L} & (2.66) \\[2em] \dfrac{dv}{dt} = \dfrac{i}{C} - \dfrac{v}{RC} & (2.67) \end{cases}$$

$$B \begin{cases} \dfrac{di}{dt} = -\dfrac{v}{L} & (2.68) \\[2em] \dfrac{dv}{dt} = \dfrac{i}{C} - \dfrac{v}{RC} & (2.69) \end{cases}$$

$$C \begin{cases} \dfrac{di}{dt} = 0 & (2.70) \\[2em] \dfrac{dv}{dt} = -\dfrac{v}{RC} & (2.71) \end{cases}$$

Network A exists during time interval t_1, network B during time t_2, and network C during the remainder of the switching cycle. It is now possible to define the duty cycle values to be used for averaging the above equations:

$$d_1 \equiv \frac{t_1}{T_s} \qquad (2.72)$$

$$d_2 \equiv \frac{t_2}{T_s} \qquad (2.73)$$

$$d_3 \equiv 1 - d_1 - d_2. \qquad (2.74)$$

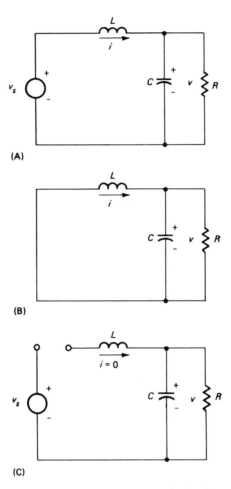

Fig. 2.19. Equivalent circuits for the buck converter in the discontinuous mode.

Before proceeding with the averaging step, a look at an additional circuit constraint is necessary because of the discontinuous conduction nature of the SPC.

For a converter operating in the continuous mode, at the beginning of each switching cycle, $I_L = i_L(0)$, and, at the end of each cycle, $I_L = i_L(T_s)$. In general, $i_L(0)$ and $i_L(T_s)$ do not equal zero. If we perturb i_L during the switching cycle, then $i_L(0) \neq i_L(T_s)$. As a consequence, the *average* value for the rate of change of inductor current *over one cycle does not equal zero* and, must be retained in the circuit equations under continuous current flow conditions.

However, in the discontinuous mode, the very definition of this operating mode requires that:

$$i_L(0) = i_L(T_s) = 0. \tag{2.75}$$

Eq. (2.75) implies that any upward rate of change of inductor current must be exactly compensated for by a corresponding downward rate of change so that the average current magnitude over one complete cycle becomes:

$$\left\langle \frac{di}{dt} \right\rangle = 0. \tag{2.76}$$

The above relationship holds even if i_L is perturbed, as can be seen in Fig. 2.20, where t_1 is extended by Δt. The average value of the perturbation must still be zero.

If we now average Eqs. (2.66) through (2.71), using the definitions given in Eqs. (2.72) through (2.74), and keeping in mind the condition stated by Eq. (2.76), the following results are obtained:

$$0 = -(d_1 + d_2)v + d_1 v_s \tag{2.77}$$

$$\frac{dv}{dt} = \left(\frac{d_1 + d_2}{C}\right) i - \frac{v}{RC}. \tag{2.78}$$

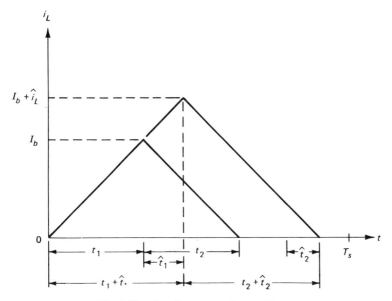

Fig. 2.20. The effect on i_L of perturbing t_1.

Once again, a mathematical model has been derived consisting of a single set of differential equations which can be easily manipulated using conventional linear circuit theory.

We can now form a circuit model using Eqs. (2.77) and (2.78), and the result is shown in Fig. 2.21A. However, the utility of this model can be improved by modifying it slightly. In Fig. 2.21B, we employ an ideal transformer to isolate v and i as separate variables. In Fig. 2.21C, the dependent generators of Fig. 2.21B are replaced with another ideal transformer. For most design purposes, the current (i) circulating in the inner loop is of no particular interest. Therefore, the two transformers of Figs. 2.21B and 2.21C can be combined to arrive at the relatively simple circuit model of Fig. 2.21D.

This intermediate discontinuous model suffers from the same problems as the continuous mode model at this point in the analysis. All of the externally controlled variables have not been separated and another variable (d_2) has been introduced which is a function of the circuit elements as well as d_1. For these reasons, the process of model refinement must be continued and, as before, we have the option of either manipulating the associated circuit equations and then determining the circuit model, or manipulating the circuit model followed by equation derivations. As was the case of the continuous model derivation, the former option will be pursued here.

However, before continuing with the model development process, we note that the model of Fig. 2.21D does have one very interesting feature—*the inductor has disappeared!* As a result, we can conclude that the second-order nature of the output filter network has been reduced to first order. The disappearance of L from the model is a direct consequence of the constraint stated in Eq. (2.76).

Because of the altered nature of the model, one of the state variables (di/dt) has been lost. We can replace this loss with another equation related to the average current, i, defined as:

$$i = \frac{i_b}{2} = (v_s - v)\left(\frac{d_1}{2R\tau_L}\right). \tag{2.79}$$

If we now perturb Eqs. (2.77) through (2.79) using the same conventions and assumptions employed in Section 2.4, and then separate them into DC and small-signal sets, the following relationships are obtained:

$$\text{DC} \begin{cases} 0 = -(D_1 + D_2)V + D_1 V_s & \text{(2.80)} \\[2mm] 0 = (D_1 + D_2)I - \dfrac{V}{R} & \text{(2.81)} \\[2mm] I = (V_s - V)\left[\dfrac{D_1}{2R\tau_L}\right] & \text{(2.82)} \end{cases}$$

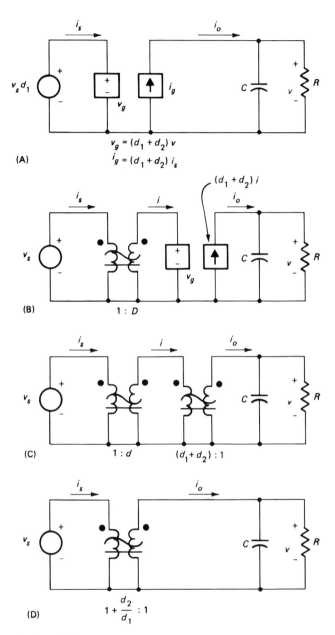

Fig. 2.21. Derivation of the state-space averaged model for the buck converter in the discontinuous mode.

$$\text{AC}\begin{cases} 0 = -(D_1 + D_2)\hat{v} + (V_s - V)\hat{d}_1 - V\hat{d}_2 + D_1\hat{v}_s & (2.83) \\\\ \dfrac{Cd\hat{v}}{dt} = (D_1 + D_2)\hat{i} - \dfrac{\hat{v}}{R} + I\hat{d}_1 + I\hat{d}_2 & (2.84) \\\\ \hat{i} = \left(\dfrac{1}{2R\tau_L}\right)[-D_1\hat{v} + (V_s - V)\hat{d}_1 + D_1\hat{v}_s]. & (2.85) \end{cases}$$

Unfortunately, the above equations still contain terms that include D_2 and $\hat{d}_2$. These can be eliminated by further manipulation of these same relationships, from which the following expressions evolve:

$$\text{DC}\begin{cases} I = \dfrac{V}{R}\sqrt{\dfrac{1 - M}{2\tau_L}} & (2.86) \\\\ V = MV_s & (2.87) \end{cases}$$

$$\text{AC}\begin{cases} \hat{i} = \left[\dfrac{M}{R\sqrt{2\tau_L(1 - M)}}\right]\hat{v} + \left(\dfrac{V}{2R\tau_L}\right)\left(\dfrac{1 - M}{M}\right)\hat{d}_1 \\\\ \qquad\qquad\qquad\qquad + \left[\dfrac{M}{R\sqrt{2\tau_L(1 - M)}}\right]\hat{v}_s \quad (2.88) \\\\ C\dfrac{d\hat{v}}{dt} = \left(\dfrac{1}{R}\right)\left(\dfrac{M - 2}{1 - M}\right)\hat{v} + \left[\left(\dfrac{V}{R}\right)\left(\dfrac{2}{M}\right)\sqrt{\dfrac{1 - M}{2\tau_L}}\right]\hat{d}_1 \\\\ \qquad\qquad\qquad\qquad + \left[\left(\dfrac{M}{R}\right)\left(\dfrac{2 - M}{1 - M}\right)\right]\hat{v}_s \quad (2.89) \end{cases}$$

$$\hat{d}_2 = \left(-\dfrac{1}{V}\sqrt{\dfrac{2\tau_L}{1 - M}}\right)\hat{v} + \left(\dfrac{1 - M}{M}\right)\hat{d}_1 + \left(\dfrac{M}{V}\sqrt{\dfrac{2\tau_L}{1 - M}}\right)\hat{v}_s \qquad (2.90)$$

These equations are defined in terms of M. If it is desirable to state them in terms of D_1, then the following relationship from Table 2.1 can be substituted:

$$M = \dfrac{2}{1 + \sqrt{1 + 8\tau_L/D_1^2}} \qquad (2.91)$$

Equations (2.86) through (2.90) provide the basis for a mathematical model. From them, one can now proceed to find an equivalent circuit. Since the parameters I and $\hat{i}$ are not of any direct interest for most SPC designs, we will develop

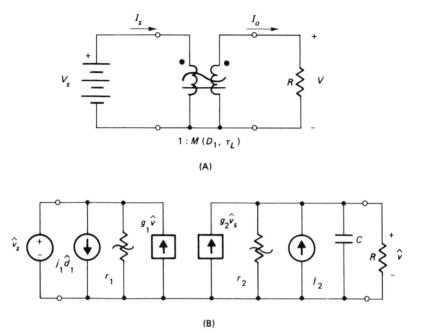

Fig. 2.22. The final state-space averaged models for the buck converter in the discontinuous mode.

a circuit model which does not include these two terms, much in the same manner as we followed to generate the circuit model in Fig. 2.21D.

The DC circuit model corresponding to the conditions of Eq. (2.87) is shown in Fig. 2.22A.

For convenience, Eq. 2.84 can be placed into the complex frequency domain by using Laplace transformation methods. Eliminating I and $\hat{i}$ and transforming this equation, we get:

$$\hat{v}_s(s) = \frac{1}{M}\left[sRC\left(\frac{1-M}{2-M}\right) + 1 \right] \hat{v}(s) - \left[\frac{2V}{M^2}\left(\frac{1}{2-M}\right) \sqrt{\frac{(1-M)^3}{2\tau_L}} \right] \hat{d}_1(s).$$

$$(2.92)$$

A circuit model which corresponds to Eq. 2.92 is illustrated in Fig. 2.22B, where:

$$j_1 = \frac{2V}{R}\sqrt{\frac{1-M}{2\tau_L}}$$

$$(2.93)$$

$$j_2 = \frac{2V}{MR} \sqrt{\frac{1-M}{2\tau_L}} \tag{2.94}$$

$$r_1 = R\left[\frac{1-M}{M^2}\right] \tag{2.95}$$

$$r_2 = R(1-M) \tag{2.96}$$

$$g_1 = \frac{1}{R}\left(\frac{M^2}{1-M}\right) \tag{2.97}$$

$$g_2 = \frac{M}{R}\left(\frac{2-M}{1-M}\right). \tag{2.98}$$

Note that the resistive elements for r_1 and r_2 have a sinusoidal line drawn through their symbols. This is done to indicate that they are AC resistances *only* and they do not physically exist under DC conditions.

To conclude, a mathematical and a circuit model for the buck SPC have been evolved, both of which can be used as small-signal modeling aids when the converter is operating in a discontinuous mode. These circuit models can easily be inserted into more complex power processing networks, keeping in mind the AC and DC models for the discontinuous mode are not identical and, therefore, it will be necessary to solve for AC and DC quantities separately. In practice, such model changeover is not a great analysis inconvenience. From the mathematical model described by Eq. 2.92, one can easily derive the important small-signal transfer characteristics of the buck SPC operating in a discontinuous mode of operation.

$$\frac{\hat{v}(s)}{\hat{v}_s(s)} = M\left[\frac{1}{1 + sRC\left(\dfrac{1-M}{2-M}\right)}\right] \tag{2.99}$$

$$\frac{\hat{v}(s)}{\hat{d}_1(s)} = \left[\left(\frac{2V}{M}\right)\left(\frac{1-M}{2-M}\right)\sqrt{\frac{1-M}{2\tau_L}}\right]\left[\frac{1}{1 + sRC\left(\dfrac{1-M}{2-M}\right)}\right] \tag{2.100}$$

3. The Boost Converter

In this chapter, we will examine the second of the two fundamental converter circuits, namely, the boost converter. The discussion will proceed in very much the same manner as it did in Chapter 2, but the end results will be quite different because the boost converter has very different properties from those of the buck converter. Even though the same components are used, the change in topology has a profound effect on how the converter operates.

In the case of the continuous-mode buck converter, the use of state-space averaging was admittedly an overkill because most of the characteristics could have been determined by viewing the circuit as a pulsed voltage source followed by a conventional low-pass filter. If that approach is used in the case of the boost converter, the answers derived would be wrong and would not include the movement of filter cutoff frequency as a function of SPC duty cycle and the nonminimum phase control-to-output transfer functions which are demonstrably present in real boost converters. The state-space averaging technique, however, does provide a very accurate model.

3.1. BOOST CONVERTER OPERATION

An ideal boost SPC circuit is shown in Fig. 3.1 with the waveforms for each operating mode established by the position of the SPDT switch (S) shown in Figs. 3.2 and 3.3.

During one complete switching cycle, S moves between positions A and B and the circuit topology changes from that shown in Fig. 3.1B to the network of Fig. 3.1C. At the beginning of the switching cycle, S is in position A. During time t_1, the inductor current (i_L) ramps upward from its value at $t = 0$ (I_a) towards a maximum value of I_b at $t = t_1$. At t_1, S moves to position B and the energy stored in L during time t_1 is now delivered to the output network (R and C). Consequently, i_L ramps downward during time t_2. If the converter is operating in the continuous mode, inductor current ramps down to a value of I_a at the end of the switching cycle, T_s. However, if the converter is operating in the discontinuous mode (Fig. 3.3), i_L at time t_2 is zero, since all of the stored energy

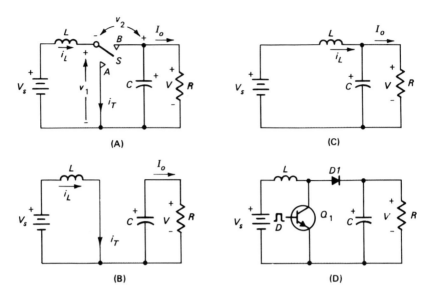

Fig. 3.1. The boost converter.

of the inductor has been delivered to the output. As was the case for the buck SPC, the operating mode of the boost SPC is of fundamental importance in determining the circuit characteristic waveforms and associated power transfer functions.

We can deduce many of the properties of the boost converter by examining the waveforms of Fig. 3.3. First, the output voltage must be greater than the input voltage. If this voltage relationship were not true, L would not discharge into the output network. In the continuous mode, the input current is non-pulsating, and its ripple magnitude can be made arbitrarily small by increasing the inductance of L. However, the input current during the discontinuous mode of operation is pulsating. The output current is always pulsating regardless of the mode of operation.

During intervals t_1 and t_3, when L is disconnected from the output, load energy needs are supported completely by C. Depending on operating mode and output power requirements, it may be necessary to make C quite large to meet the usual requirement of small output voltage ripple.

Current i_T is the collector current of the transistor switch (Q_1) in Fig. 3.1D. Comparing its waveform in Fig. 3.2 and that in Fig. 3.3, we see that its peak value is much higher in the discontinuous mode for a constant output voltage and load. It is evident that Q_1 must switch higher peak currents when it is turning off.

The DC input current (I_s) is the average value of i_L. For a given input power:

$$P_s = I_s V_s \tag{3.1}$$

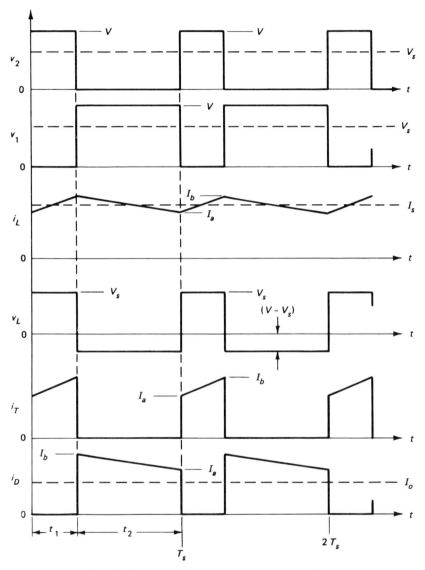

Fig. 3.2. Boost converter continuous mode waveforms.

Since I_a and I_b do not vary greatly with changes in V_s for a constant output voltage (V), the peak values for I_T and I_b in Fig. 3.3 are primarily functions of I_s. This dependency in the boost SPC has a counterpart in the buck SPC, where the peak values of I_T and I_b are determined by average output current (I_o). Therefore, in the boost SPC, as V_s increases, values of $I_{T(\mathrm{avg})}, I_{T(\mathrm{rms})}$ as well as I_b decrease.

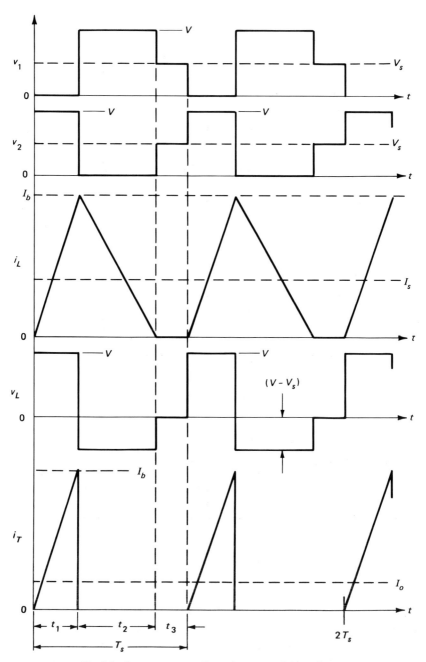

Fig. 3.3. Boost converter discontinuous mode waveforms.

3.2. LARGE-SIGNAL STEADY-STATE CHARACTERIZATION

In this section, the characterization procedure will presume the ideal boost SPC of Fig. 3.4 for simplicity. In a high efficiency converter, the parasitic resistances must be small so that their omission will have only a small effect on the accuracy of the analysis.

As shown in Fig. 3.4, an input filter network $(L1, C1)$ has been added to reflect a practical SPC. Equations defining the currents indicated in this circuit are summarized in Tables 3.1 and 3.2, where:

$$D_1 = \frac{t_1}{T_s} \tag{3.2}$$

$$D_2 = \frac{t_2}{T_s} \tag{3.3}$$

$$M = \frac{V}{V_s} \tag{3.4}$$

$$\tau_L = \frac{L}{RT_s} \tag{3.5}$$

$$I_o = \frac{V}{R} \tag{3.6}$$

$$I_s = I_{L\,(\mathrm{avg})} \tag{3.7}$$

Considerable insight can be gained into the characteristics of the boost SPC by graphing some of the equations from Tables 3.1 and 3.2; however, we must first

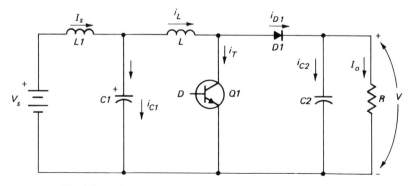

Fig. 3.4. Definition of the boost converter component currents.

**TABLE 3.1A. Definition of Variables for
Boost Converter Equations.**

τ_L	L/RT_s
T_S	Switching period
τ_{LC}	L_c/RT_s
L_c	Critical inductance
τ_{LC}	$(M - 1)/2M^3 = D_1(1 - D_1)^2/2$
D_1	ON duty cycle of switch
D_2	Diode conduction duty cycle
M	V/V_s
I_a	Minimum inductor current
I_b	Maximum inductor current

define the regions of operation where each set of equations will be valid. The boundary between the continuous and discontinuous modes is defined by τ_{LC}, where:

$$\tau_{LC} = \frac{M - 1}{2M^3} \tag{3.8}$$

or alternatively:

$$\tau_{LC} = \frac{D_1(1 - D_1)^2}{2}. \tag{3.9}$$

Graphs of Eqs. (3.8) and (3.9) are given in Figs. 3.5 and 3.6, respectively. We see that if $\tau_L > 2/27$, or approximately 0.075, then the converter will always remain in the continuous mode. Another interesting condition shown is the fact that, if $\tau_L < 0.075$, then an upper *and* a lower boundary condition exist, at which a mode change will occur. For example, if $\tau_L = 0.025$ the operating mode will change from continuous to discontinuous for $D_1 = 0.05$, and back to continuous when $D_1 = 0.73$. As τ_L is made larger, the range of D_1 over which the discontinuous mode exists will become progressively smaller.

With the areas of validity for the equations of Tables 3.1 and 3.2 established, graphical examination of one of them can now proceed. Fig. 3.7 shows the variation of M versus changes in D_1. For the continuous mode, this relationship

TABLE 3.1B. Boost Converter Equations for Continuous Inductor Current.

Dependent Variable	$f(M, \tau_L, V, R)$	$f(D_1, \tau_L, V, R)$
M	M	$\dfrac{1}{1 - D_1}$ *
D_1	$\dfrac{M - 1}{M}$ *	$D_1 = D$
D_2	$\dfrac{1}{M}$	$1 - D$
I_a	$\dfrac{V}{R}\left[M - \left(\dfrac{1}{2\tau_L}\right)\left(\dfrac{M - 1}{M^2}\right)\right]$	$\dfrac{V}{R}\left[\dfrac{1}{1 - D} - \left(\dfrac{1}{2\tau_L}\right)D(1 - D)\right]$
I_b	$\dfrac{V}{R}\left[M + \left(\dfrac{1}{2\tau_L}\right)\left(\dfrac{M - 1}{M^2}\right)\right]$	$\dfrac{V}{R}\left[\dfrac{1}{1 - D} + \left(\dfrac{1}{2\tau_L}\right)D(1 - D)\right]$
$I_{L\,(avg)}$	$\dfrac{V}{R}M$	$\left(\dfrac{V}{R}\right)\left(\dfrac{1}{1 - D}\right)$
$I_{L\,(rms)}$	$\dfrac{V}{R}\left[M^2 + \dfrac{1}{3}\left(\dfrac{1}{2\tau_L}\right)^2\left(\dfrac{M - 1}{M}\right)^2\right]^{1/2}$	$\dfrac{V}{R}\left[\left(\dfrac{1}{1 - D}\right)^2 + \dfrac{1}{3}\left(\dfrac{1}{2\tau_L}\right)^2 D(1 - D)\right]^{1/2}$
$I_{C1\,(rms)}$	$\dfrac{V}{R}\left(\dfrac{1}{12\tau_L}\right)\left(\dfrac{M - 1}{M^2}\right)$	$\dfrac{V}{R}\left[\dfrac{D(1 - D)}{12\tau_L}\right]$
$I_{C2\,(rms)}$	$\dfrac{V}{R}\left[(M - 1) + \dfrac{1}{3}\left(\dfrac{1}{2\tau_L}\right)^2\left(\dfrac{1}{M}\right)\left(\dfrac{M - 1}{M^2}\right)^2\right]^{1/2}$	$\dfrac{V}{R}\left[\dfrac{D}{1 - D} + \dfrac{D}{12}\left(\dfrac{1 - D}{\tau_L}\right)^2\right]^{1/2}$
$I_{T\,(avg)}$	$\dfrac{V}{R}(M - 1)$	$\left(\dfrac{V}{R}\right)\left(\dfrac{D}{1 - D}\right)$
$I_{T\,(rms)}$	$\dfrac{V}{R}\left[M(M - 1) + \dfrac{1}{3}\left(\dfrac{1}{2\tau_L}\right)^2\left(\dfrac{M - 1}{M^2}\right)^3\right]^{1/2}$	$\dfrac{V}{R}\left[\dfrac{D}{(1 - D)^2} + \dfrac{1}{3}\left(\dfrac{1}{2\tau_L}\right)^2 D^3(1 - D)^3\right]^{1/2}$
$I_{D1\,(avg)}$	$\dfrac{V}{R}$	$\dfrac{V}{R}$

*For $M > 3$, see equation 3.34 in text.

is quite nonlinear, with the value M rising rapidly as D_1 is increased. In contrast, in the discontinuous mode, the graphical relationship between M and D_1 is very nearly a straight line. This latter situation is surprising when one looks casually at the defining equation of this graph:

$$M = \frac{\sqrt{1 + 2D_1^2/\tau_L}}{2} x + \frac{1}{2} \tag{3.10}$$

TABLE 3.2A.   Boost Converter Equations for Discontinuous Inductor Current.

Dependent Variable	$f(M, \tau_L, V, R)$	$f(D_1, \tau_L, V, R)$
M	M	$\dfrac{1 + \sqrt{1 + 2D_1^2/\tau_L}}{2}$
D_1	$\sqrt{2\tau_L M(M-1)}$	D_1
D_2	$\sqrt{\dfrac{2\tau_L M}{M-1}}$	$\left(\dfrac{\tau_L}{D_1}\right)(1 + \sqrt{1 + 2D_1^2/\tau_L})$
I_a	0	0
I_b	$\dfrac{V}{R}\sqrt{\dfrac{2(M-1)}{\tau_L M}}$	$\dfrac{V}{R}\left[\dfrac{\sqrt{1 + 2D_1^2/\tau_L} - 1}{D_1}\right]$
$I_{L(\text{avg})}$	$\dfrac{V}{R}M$	$\dfrac{V}{R}\left(\dfrac{1 + \sqrt{1 + 2D_1^2/\tau_L}}{2}\right)$
$I_{L(\text{rms})}$	$\dfrac{V}{R}\left[\dfrac{8}{9}\dfrac{M(M-1)}{\tau_L}\right]^{1/4}$	$\dfrac{V}{R}\left[\dfrac{2}{3}\dfrac{D_1}{\tau_L}\right]^{1/2}$
$I_{C1(\text{rms})}$	$\dfrac{V}{R}\left[\sqrt{\dfrac{8}{9}\dfrac{M(M-1)}{\tau_L}} - M^2\right]^{1/2}$	$\dfrac{V}{R}\left[\dfrac{1 + \sqrt{1 + 2D_1^2/\tau_L}}{2}\right]\left[\sqrt{\dfrac{8}{9}\dfrac{D_1(1-D_1)^2}{\tau_L}} - 1\right]^{1/2}$
$I_{C2(\text{rms})}$	$\dfrac{V}{R}\left[\sqrt{\dfrac{8}{9}\dfrac{M-1}{\tau_L M}} - 1\right]^{1/2}$	$\dfrac{V}{R}\left[\dfrac{2}{3}\left(\dfrac{\sqrt{1 + 2D_1^2/\tau_L} - 1}{D_1}\right) - 1\right]^{1/2}$
$I_{T(\text{avg})}$	$\dfrac{V}{R}(M-1)$	$\dfrac{V}{R}\left[\dfrac{\sqrt{1 + 2D_1^2/\tau_L} - 1}{2}\right]$
$I_{T(\text{rms})}$	$\dfrac{V}{R}\left[\dfrac{8}{9}\dfrac{(M-1)^3}{\tau_L M}\right]^{1/4}$	$\dfrac{V}{R}\left[\dfrac{\sqrt{1 + 2D_1^2/\tau_L} - 1}{\sqrt{3D_1}}\right]$
$I_{D1(\text{avg})}$	$\dfrac{V}{R}$	$\dfrac{V}{R}$

However, in most of the area where Eq. (3.10) is valid, we can state:

$$\frac{2D_1^2}{\tau_L} \gg 1. \tag{3.11}$$

Using the condition of Eq. (3.11) the value of M set by Eq. (3.10) can be approximated as:

$$M \approx \frac{1}{2} + \frac{D_1}{\sqrt{2\tau_L}}. \tag{3.12}$$

TABLE 3.2B. Equations Normalized to τ_{LC}.

$$\frac{I_a}{I_s} = 1 - \frac{\tau_{LC}}{\tau_L}$$

$$\frac{I_b}{I_s} = 1 + \frac{\tau_{LC}}{\tau_L}$$

$$\frac{I_{L\,(\text{rms})}}{I_s} = \left[1 + \frac{1}{3}\left(\frac{\tau_{LC}}{\tau_L}\right)^2\right]^{1/2}$$

Continuous Mode

$$\frac{I_b}{I_s} = 2\sqrt{\frac{\tau_{LC}}{\tau_L}}$$

$$\frac{I_{L\,(\text{rms})}}{I_s} = \left[\frac{16}{9}\left(\frac{\tau_{LC}}{\tau_L}\right)\right]^{1/4}$$

Discontinuous Mode

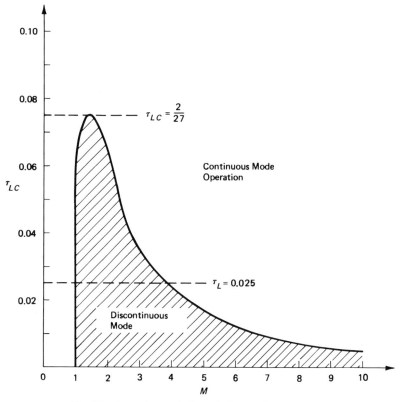

Fig. 3.5. Operating mode boundaries as a function of M.

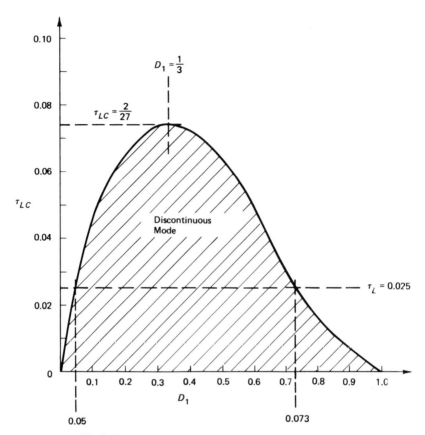

Fig. 3.6. Operating mode boundaries as a function of D_1.

If we compare the results of Eqs. (3.10) and (3.12) for values of D_1 and τ_L set at 0.5 and 0.01, respectively, an error of less than 1% will be found. Therefore, the simpler form of Eq. (3.10) will give results accurate for most analysis needs. Even if the value of τ_L is increased by a factor of ten, the error introduced by the use of Eq. (3.10) is only 3%.

As was the case for the buck SPC of the last chapter, the voltage control characteristics of Fig. 3.7 differs greatly between the two conduction modes. Although the conditions of this graph apply to an ideal boost SPC, the graph is a good representation of the control characteristic for practical boost converters, where M is less than three. For continuous-mode M values larger than three, the effects of any component parasitic resistances become quite noticeable. The associated plot of Fig. 3.7 does not continue its exponential climb, but actually reaches a peak and then begins to fall. This change will be demonstrated in Section 3.4, when component parasitics are added to the ideal boost SPC model.

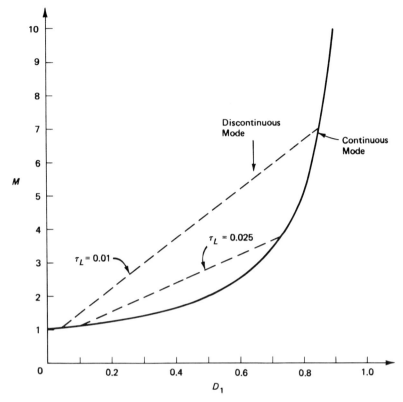

Fig. 3.7. DC control characteristic for the boost converter.

The variations in I_b and $I_{L\,(rms)}$ as τ_L is varied are shown in Figs. 3.8 and 3.9, respectively. From these graphs, the effect of τ_L on the amplitude of these particular currents can be clearly seen. In very much the same manner as was observed in the buck SPC, the amplitudes of these currents are always higher in the discontinuous mode and, in some cases, they can be very large. Thus, the use of the discontinuous mode for boost SPC operation is not advantageous for high output power.

3.3. LARGE-SIGNAL DYNAMIC CHARACTERIZATION

If the output load of a boost SPC is removed, or short-circuited, the converter reacts in a manner unlike that observed in the buck SPC of Chapter 2 under similar conditions.

In an output short-circuit condition, reducing the duty cycle of the switch to zero will not limit the amount of current drawn from the source, because the

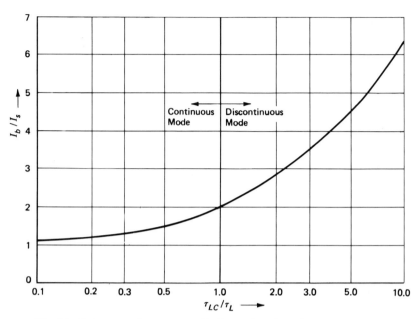

Fig. 3.8. Inductor peak current amplitude as a function of inductance value.

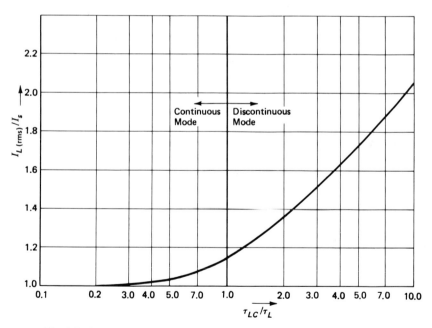

Fig. 3.9. Inductor rms current amplitude as a function of inductance value.

switch is *not* in series with it. Reducing the duty cycle of switch will provide output overload protection only until the output voltage equals the source voltage. At that point, there is no inherent mechanism for output current limiting, other than parasitic inductor resistance and diode impedance, both of which are normally very small. If output current overload must be tolerated, then an additional switch in series with the source must be added. This is a fundamental problem common to all boost-derived converters. Fortunately, the problem can be resolved fairly simply. Practical solutions are treated in Chapter 9.

If the boost SPC output is open-circuited, another serious problem arises. If the switching action of the SPC is continued in the absence of an output load, the energy stored in C will increase. The output voltage will rise until a component fails.

A short anecdote will serve to emphasize the potential seriousness of the problem. Some years ago, one of the authors was familiarizing a young electronics engineer with the design of SPCs. One of his job assignments was to design and build a basic boost SPC, and he was forewarned never to operate the boost SPC without proper loading. This admonition apparently was not taken seriously, for shortly afterwards, a loud bang was heard at the engineer's work bench. Upon investigation by co-workers, he was found to be bleeding from his forehead, almost exactly between the eyes, where the inside portions of an electrolytic capacitor had struck him. Apparently, the output filter capacitor had exploded when the load of the converter was removed. He finally got the point (no pun intended) and, we trust you do too, from this account of a very unfortunate experience.

Applications where the output loading of a boost SPC could be removed are frequently encountered by designers. In these instances, overvoltage or stress problems from unloading can be eliminated by adding auxiliary circuitry in the SPC control system to sense an overvoltage condition and then reduce switch duty cycle D to zero.

When V_s is first applied to the boost SPC of Fig. 3.1D, and the shunt switch is OFF, the input inrush current is limited only by the characteristics of a low-pass filter network formed by R, L, and C. The shape of the inrush source current transient is one-half cycle of a damped sinusoid, as shown in Fig. 3.10. The maximum value of this transient may be many times that of the steady-state average value of I_s. This high peak current can produce damage to filter elements or to the source. For these reasons, SPC input specifications often call for limiting the amplitude of inrush current.

An additional complication may arise if the inrush current reaches a value large enough to cause the core of L to saturate. Should this happen, the waveforms in Fig. 3.10 will no longer apply, and even higher values of inrush current will occur, limited only by source impedances and parasitic resistances of L, D, and C.

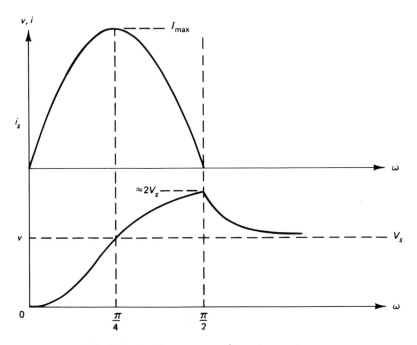

Fig. 3.10. Inrush current waveform at source turn-on.

In addition to potentially high inrush currents at turnon, there is another problem peculiar to the boost SPC relative to starting sequence. Some boost designs are structured so as to begin SPC switching action with the application of V_s. When V_s is first applied, the output voltage (V_o) will be initially zero and begin to rise toward the desired output voltage value. During this build-up interval, the auxiliary voltage control circuitry will push the value of D to a maximum and decrease it as output voltage increases. If the maximum value of D is too large, the circuit will lock up and V_o will not be built up. This situation is usually accompanied by both excessive input inductor as well as switch current values. It is normal practice, therefore, to limit the maximum value of D to a value slightly larger than that required for normal output voltage. An analytical determination of the maximum duty cycle is presented in Section 3.4.

It is well to remember that these potentially damage-producing properties of the basic boost SPC are retained in their more complex derivative topologies. In boost-derived SPCs which contain a DC transformation element, the inrush current occurs at the onset of SPC switching action, rather than at the time when V_s is first applied. Nevertheless, the potential stress problems discussed earlier are still possible.

3.4. STATE-SPACE AVERAGED MODEL FOR THE CONTINUOUS MODE

The same processes that were used in Chapter 2 for deriving general small-signal models for the buck SPC can be repeated here for the boost SPC.

Figure 3.11 shows the equivalent circuits for each of the two possible states of the boost SPC in the continuous mode. We have now included the parasitic resistances of the inductor and output capacitor to make our models practical in nature.

The corresponding circuit equations, in state-variable form, for the network of Fig. 3.11A are:

$$\frac{di}{dt} = -\left[\frac{R_L}{L}\right]i + \frac{v_s}{L} \tag{3.13}$$

$$\frac{dv}{dt} = -\frac{v}{(R + R_c)C} \tag{3.14}$$

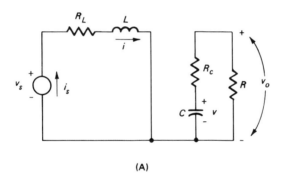

(A)

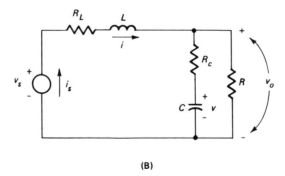

(B)

Fig. 3.11. Boost converter equivalent circuits.

$$v_o = \left[\frac{R}{R + R_c}\right] v \tag{3.15}$$

$$i_s = i. \tag{3.16}$$

For network B of Fig. 3.11, the describing circuit equations are:

$$\frac{di}{dt} = -\left[\frac{R_L + (R\|R_c)}{L}\right] i - \frac{1}{L}\left[\frac{R}{R + R_c}\right] v + \frac{v_s}{L} \tag{3.17}$$

$$\frac{dv}{dt} = \left[\frac{R}{(R + R_c)C}\right] i - \frac{v}{(R + R_c)C} \tag{3.18}$$

$$v_o = (R\|R_c)i + \left[\frac{R}{R + R_c}\right] v \tag{3.19}$$

$$i_s = i. \tag{3.20}$$

Using the same assumptions for averaging that were established in Chapter 2, the state-space averaged circuit equations for the boost SPC are:

$$\frac{di}{dt} = -\left[\frac{R_L + (1 - d)(R\|R_c)}{L}\right] i - \left[\frac{(1 - d)R}{L(R + R_c)}\right] v + \frac{v_s}{L} \tag{3.21}$$

$$\frac{dv}{dt} = \left[\frac{(1 - d)R}{(R + R_c)C}\right] i - \frac{v}{(R + R_c)C} \tag{3.22}$$

$$v_o = (1 - d)(R\|R_c)i + \left[\frac{R}{R + R_c}\right] v \tag{3.23}$$

$$i_s = i. \tag{3.24}$$

A circuit model corresponding to the above equations is shown in Fig. 3.12. In this model, an output low-pass filter network is evident as was the case for the buck SPC model in the continuous mode. However, the values for L and its series resistances are now modulated functions of the switch duty cycle. It follows then that the output filter network characteristics will change as a function of d. This was not the case in the buck SPC model and represents an important difference between these two converters.

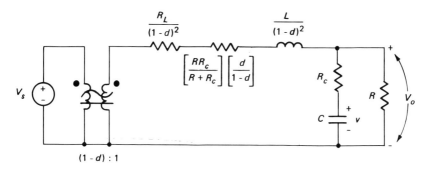

Fig. 3.12. Initial state-space averaged model for the boost converter.

Continuing the process of model development, we can now perturb Eqs. (3.21) through (3.24) using the same small-signal conventions that were employed in Chapter 2. As a result of these perturbations, the following relationships are established:

DC
$$0 = -[R_L + (R\|R_c)D']I - \left(\frac{RD'}{R + R_c}\right)V + V_s \tag{3.25}$$

$$0 = (RD')I - V \tag{3.26}$$

$$V_o = (R\|R_c)D'I + \left(\frac{R}{R + R_c}\right)V \tag{3.27}$$

$$I_s = I \tag{3.28}$$

AC
$$\frac{d\hat{i}}{dt} = -\left[\frac{R_L + (R\|R_c)D'}{L}\right]\hat{i} - \left[\frac{RD'}{(R + R_c)L}\right]\hat{v}$$
$$+ \left(\frac{V_o}{D'L}\right)\left(\frac{D'R + R_c}{R + R_c}\right)\hat{d} + \frac{\hat{v}_s}{L} \tag{3.29}$$

$$\frac{d\hat{v}}{dt} = \left[\frac{RD'}{(R + R_c)C}\right]\hat{i} - \left[\frac{I}{(R + R_c)C}\right]\hat{v} - \left[\frac{V_o}{D'(R + R_c)C}\right]\hat{d}, \qquad \hat{i}_s = \hat{i} \tag{3.30}$$

$$\hat{v}_o = (R\|R_c)D'\hat{i} + \left(\frac{R}{R + R_c}\right)\hat{v} - \left[\frac{V_o R_c}{D'(R + R_c)}\right]\hat{d}. \tag{3.31}$$

From the DC equations, we can readily derive expressions for V, I, and M as:

$$I = \frac{V_o}{D'R'}, \qquad D' = 1 - D \tag{3.32}$$

$$V = V_o \tag{3.33}$$

$$M = \frac{D'R}{R'} \tag{3.34}$$

where

$$R' = R_L + (R \| R_c)D' + \frac{R^2 (D')^2}{R + R_c}. \tag{3.35}$$

Equation (3.34) can be rearranged so that it takes the form of the voltage gain function of an ideal boost SPC multiplied by a correction factor produced by the SPC's parasitic resistances:

$$M = \left(\frac{1}{D'}\right)\left[\frac{(D')^2 R}{R'}\right]. \tag{3.36}$$
$$\uparrow \qquad \uparrow$$
$$\text{IDEAL} \quad \text{CORRECTION FACTOR}$$

Figure 3.13 shows a graph of Eq. (3.34) comparing an ideal M versus D characteristic ($R_L = R_c = 0$) to one in which $R_L = R_c = 0.01R$. As shown, for M values less than 3, the difference between actual and ideal relationships is very small. For M values greater than 3, the two curves diverge from one another rapidly and the effect of the correction factor in Eq. (3.36) becomes apparent in the nonideal case. Values of parasitic resistances for an efficient conversion process are usually less than 1% of the converter's maximum load value, and, frequently, even smaller parasitic values can be realized by proper filter component selection.

The departure of Eq. (3.36) from the ideal curve of Fig. 3.13 at high duty cycles poses a problem in SPC applications. In this example, for $R_L = R_c = 0.01R$, the maximum value for M is 4.7 and occurs at a duty cycle of 0.88. Should the source voltage (V_s) drop in value, any output voltage regulation control electronics of the converter would attempt to increase D, and therefore M, to maintain a constant output voltage. If the SPC's regulation controls were designed so as to allow a duty cycle of 0.88 to be exceeded under these circumstances, then the output voltage would drop, resulting in a further increase in the value of D. This increase in D would continue until a maximum control limit value is

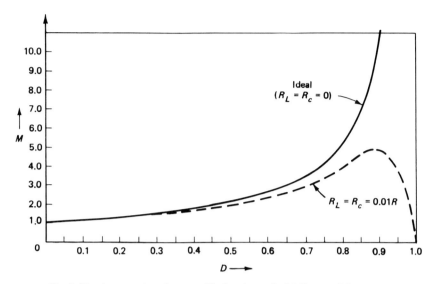

Fig. 3.13. A comparison between ideal and practical DC control functions.

reached. Therefore, in our practical example, it is possible for related voltage control electronics to produce a *latch-up* condition, wherein the SPC switch would be placed in a full conduction condition. This is a primary reason that all boost SPC systems should be designed so that their maximum duty cycle is limited to prevent latch-up. For a more detailed discussion of this lock-up problem, the reader should examine the contents of Ref. 4 given in the text bibliography.

We can now complete our mathematical model by transforming Eqs. (3.28) through (3.31) into the complex frequency domain:

$$\frac{V_o}{D'}\left[\frac{D'R + R_c}{R + R_c}\right]\hat{d}(s) + \hat{v}_s(s) = [sL + R_L + D'(R\|R_c)]\,\hat{i}(s) + \left[\frac{RD'}{R + R_c}\right]\hat{v}_s$$

$$(3.37)$$

$$\left(\frac{V_o}{D'}\right)\hat{d}(s) = (D'R)\hat{i}(s) - [1 + s(R + R_c)C]\hat{v}(s) \qquad (3.38)$$

$$\hat{v}_o(s) = D'(R\|R_c)\hat{i}(s) + \left[\frac{R}{R + R_c}\right]\hat{v}(s)$$

$$-\left[\frac{V_o R_c}{D'(R + R_c)}\right]d(s) \qquad (3.39)$$

$$\hat{i}_s(s) = \hat{i}(s). \tag{3.40}$$

Examination of the mathematical model equations shows that they have exactly the same form as the model for the buck converter derived in Chapter 2. Only the proportionality constants produced by the circuit element values are different. This implies that any equivalent circuit model topology established by the above equations will be *identical* to that of the buck SPC, and differ only in element values. From this observation and a little bit of equation manipulation (see Chapter 10), we can formulate the small-signal continuous circuit model of the boost SPC as shown in Fig. 3.14, where:

$$E = V_o \left[\frac{R}{R + R_c} - \frac{R_L}{(D')^2 R} \right] \tag{3.41}$$

$$f_1(s) = 1 - s \left[\frac{L}{\dfrac{(D'R)^2}{R + R_c} - R_L} \right] \tag{3.42}$$

$$J = \frac{V_o}{(D')^2 R}, \qquad f_2(s) = 1 \tag{3.43}$$

$$R_e = \frac{R_L + (R \| R_c) D D'}{(D')^2} \tag{3.44}$$

$$L_e = \frac{L}{(D')^2}, \qquad f_c = \frac{1}{2\pi\sqrt{L_e C}}. \tag{3.45}$$

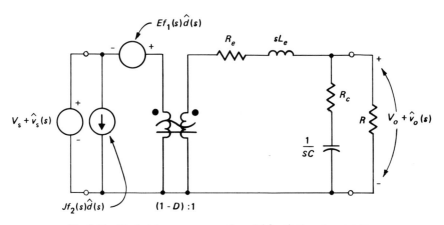

Fig. 3.14. Final state-space averaged model for the boost converter.

Once again, the output network of the model in Fig. 3.14 takes the form of a low-pass filter; however, some filter elements (R_e and L_e) are functions of D. In a closed-loop voltage control application, these variations will cause f_c to change as a function of D. Obviously, this will complicate the design of the SPC voltage control electronics from stability and bandwidth standpoints.

We see also that the $\hat{d}(s)$ voltage generator has a complex amplitude $f_1(s)$ and that $f_1(s)$ has a *right* half-plane zero! The right half-plane zero appears as a consequence of the switching action, and seriously complicates the problem of stabilizing the control loop.

A simple physical explanation for the presence of this zero can be found by considering boost control circuit reaction to a step increase in $D(\Delta D)$, which is equivalent to commanding a step increase in output voltage. The initial reaction of the boost SPC system would be to keep its shunt switch on for a corresponding period, $D + \Delta D$. During this interval, V_o will drop slightly, since the output voltage must be sustained by the energy in $C2$ for a slightly longer time than was necessary on the immediate preceeding switching cycle of the SPC. The initial slope of the output response (dv_o/dt) is therefore negative for a positive input change. This phenomenon is characteristic of a system with a zero in the right half of the complex frequency plane. The negative rate of change of v_o exists only momentarily, and then becomes positive in value as output voltage rises in response to control increases in converter duty cycle D.

From the small-signal equations, we can now derive some SPC transfer functions of interest:

$$\frac{\hat{v}_o(s)}{\hat{v}_s(s)} = \left[\frac{D'R}{R'}\right] \cdot \left\{ \frac{(1 + sR_cC)}{s^2LC\left(\dfrac{R + R_c}{R'}\right) + s\left[\dfrac{L}{R'} + \left(\dfrac{RR_L + R_cR_L + D'RR_c}{R'}\right)C\right] + 1} \right\}$$

$$(3.46)$$

$$\frac{\hat{v}_o(s)}{\hat{d}(s)} = \left[\frac{V_o}{D'R'}\right]\left[\frac{(D'R')^2}{R + R_c} - R_L\right]$$

$$\cdot \left[\frac{(1 + sR_cC)\left[1 - s\left(\dfrac{L}{\dfrac{(D'R)^2}{R + R_c} - R_L}\right)\right]}{s^2LC\left[\dfrac{R + R_c}{R'}\right] + s\left[\dfrac{L}{R'} + \left(\dfrac{RR_L + R_cR_L + D'RR_c}{R'}\right)C\right] + 1}\right] \quad (3.47)$$

In both of these equations, a zero appears because of the presence of output capacitor ESR (R_c) as well as a complex pole that moves as a function of D.

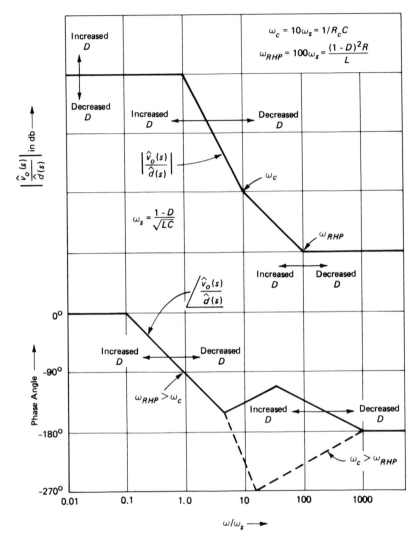

Fig. 3.15. Asymptotic gain and phase approximations for the small-signal control-to-output transfer function of a typical boost converter.

However, the right half-plane zero discussed earlier appears only in the control-to-output transfer function. An asymptotic gain and phase approximation for this latter function is shown in Fig. 3.15, where the deleterious effect of the right half-plane zero on system stability is very apparent when compared to the same graph for the buck SPC (Fig. 2.17). In Fig. 3.15, it has been presumed that the ESR effect of the output capacitor results in a break frequency lower than that caused by the right half-plane zero. If the frequency relationships of these two zero break points are interchanged, the gain plot will remain unchanged, but the phase angle plot (as shown by the dashed line) will move in a direction so as to even more severely complicate control measures for system stability.

3.5. STATE-SPACE AVERAGED MODEL FOR THE DISCONTINUOUS MODE

Like its counterpart SPC of Chapter 2, the boost converter assumes three different states during the discontinuous mode of operation. These are shown in Fig. 3.16. For each of the networks of Fig. 3.16, a set of equations can be written:

$$A \begin{cases} \dfrac{di}{dt} = \dfrac{v_s}{L} & (3.48) \\[2em] \dfrac{dv}{dt} = -\dfrac{v}{RC} & (3.49) \end{cases}$$

$$B \begin{cases} \dfrac{di}{dt} = -\dfrac{v}{L} + \dfrac{v_s}{L} & (3.50) \\[2em] \dfrac{dv}{dt} = \dfrac{i}{C} - \dfrac{v}{RC} & (3.51) \end{cases}$$

$$C \begin{cases} \dfrac{di}{dt} = 0 & (3.52) \\[2em] \dfrac{dv}{dt} = -\dfrac{v}{RC}. & (3.53) \end{cases}$$

The duty cycle periods associated with the three networks of Fig. 3.16 are defined as follows:

$$A: \quad d_1 = \frac{t_1}{T_s} \tag{3.54}$$

$$B: \quad d_2 = \frac{t_2}{T_s} \tag{3.55}$$

$$C: \quad d_3 = 1 - d_1 - d_2. \tag{3.56}$$

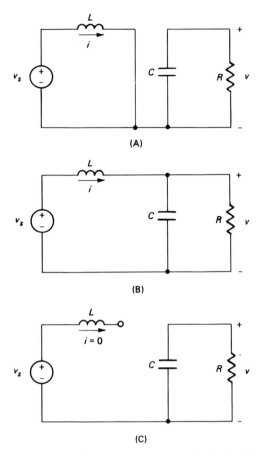

Fig. 3.16. Equivalent circuits for the boost converter operating in the discontinuous mode.

We can now perform the state-space averaging step, keeping in mind that di/dt is zero for the same reasons that were evident in the buck SPC during the discontinuous mode. The resulting averaged equations are:

$$0 = -d_2 v + (d_1 + d_2)v_s \tag{3.57}$$

$$\frac{dv}{dt} = \left(\frac{d_2}{C}\right) i - \frac{v}{RC}. \tag{3.58}$$

Again, the lost state variable, i, can be replaced by:

$$i \equiv \frac{i_b}{2} = \left[\frac{d_1}{2R\tau_L}\right]v_s. \tag{3.59}$$

Construction of the circuit topology corresponding to Eqs. (3.57), (3.58), and (3.59) would reveal that the inductor does not appear in the resulting circuit.

Perturbation of Eqs. (3.57) through (3.59) and separation of AC and DC components result in the following expressions:

DC $\left\{ \begin{array}{l} \quad \\ \quad \\ \quad \\ \quad \end{array} \right.$

$$0 = -D_2 V + (D_1 + D_2)V_s \tag{3.60}$$

$$0 = D_2 I - \frac{V}{R} \tag{3.61}$$

$$I = \frac{D_1 V_s}{2R\tau_L} \tag{3.62}$$

AC $\left\{ \begin{array}{l} \quad \\ \quad \\ \quad \\ \quad \end{array} \right.$

$$0 = -D_2 \hat{v} + V_s \hat{d}_1 - (V - V_s)\hat{d}_2 + (D_1 + D_2)\hat{v}_s \tag{3.63}$$

$$\frac{dv}{dt} = \left[\frac{D_2}{C}\right]\hat{i} - \frac{\hat{v}}{RC} + \left[\frac{I}{C}\right]\hat{d}_2 \tag{3.64}$$

$$\hat{i} = \left[\frac{V_s}{2R\tau_L}\right]\hat{d}_1 + \left[\frac{D_1}{2R\tau_L}\right]\hat{v}_s \tag{3.65}$$

The mathematical model described by the above relationships has terms including the dependent variables D_2 as well as $\hat{d}_2$. With a bit of manipulation, these dependent terms can be eliminated and the model then becomes:

DC $\left\{ \begin{array}{l} \quad \\ \quad \end{array} \right.$

$$I = \frac{V}{R}\sqrt{\frac{M-1}{2\tau_L M}} \tag{3.66}$$

$$V = MV_s \tag{3.67}$$

AC $\left\{ \begin{array}{l} \quad \\ \quad \\ \quad \\ \quad \\ \quad \\ \quad \end{array} \right.$

$$\hat{i} = \frac{V}{R}\left[\frac{1}{2\tau_L M}\right]\hat{d}_1 + \frac{1}{R}\left[\sqrt{\frac{M(M-1)}{2\tau_L}}\right]\hat{v}_s \tag{3.68}$$

$$C\frac{dv}{dt} = -\frac{1}{R}\left[\frac{2M-1}{M-1}\right]\hat{v} + \left[\frac{2V}{R}\right]\frac{\hat{d}_1}{\sqrt{2\tau_L M(M-1)}} + \frac{M}{R}\left[\frac{2M-1}{M-1}\right]\hat{v}_s \tag{3.69}$$

$$\hat{d}_2 = -\frac{1}{V}\left[\sqrt{\frac{2\tau_L M^3}{(M-1)^3}}\right]\hat{v} + \frac{\hat{d}_1}{M-1} + \frac{1}{V}\left[\frac{M}{M-1}\right]^2 [\sqrt{2\tau_L M(M-1)}]\hat{v}_s. \tag{3.70}$$

The above equations are stated in terms of M. If one would rather have them expressed in terms of D_1, then a substitution is possible:

$$M = \frac{1 + \sqrt{1 + 2D_1^2/\tau_L}}{2}. \tag{3.71}$$

For convenience, we will transform Eq. (3.69) into the complex frequency domain, obtaining:

$$M\hat{v}_s(s) = \left[1 + sRC\left(\frac{M-1}{2M-1}\right)\right]\hat{v}(s) - \left[\frac{2V}{2M-1}\right]\left[V\sqrt{\frac{M-1}{2\tau_L M}}\right]\hat{d}_1. \tag{3.72}$$

If we compare Eqs. (3.67) and (3.72) with their buck SPC counterparts of Chapter 2 in the discontinuous mode [Eqs. (2.87) and (2.92), respectively], the equation form is identical and each set differs only in the proportionality constant. This allows the use of the buck circuit model with different element values, as shown in Fig. 3.17, where:

$$j_1 = \frac{2V}{R}\sqrt{\frac{M}{2\tau_L(M-1)}} \tag{3.73}$$

$$j_2 = \frac{2V}{R\sqrt{2\tau_L M(M-1)}} \tag{3.74}$$

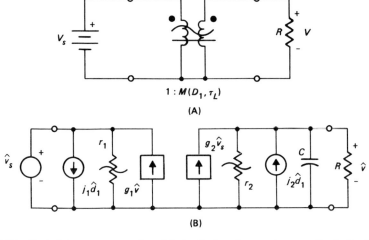

$$1 : M(D_1, \tau_L)$$

(A)

(B)

Fig. 3.17. Final state-space averaged models for the boost converter operating in the discontinuous mode.

$$r_1 = \left[\frac{M-1}{M^3}\right]R \qquad (3.75)$$

$$r_2 = \left[\frac{M-1}{M}\right]R \qquad (3.76)$$

$$g_1 = \frac{1}{R}\left[\frac{M}{M-1}\right] \qquad (3.77)$$

$$g_2 = \frac{1}{R}\left[\frac{M(2M-1)}{M-1}\right]. \qquad (3.78)$$

From the models of Figs. 3.17A and 3.17B, the input-to-output and control-to-output transfer functions can easily be derived as:

$$\frac{\hat{v}(s)}{\hat{v}_s(s)} = \frac{M}{1 + sRC\left(\dfrac{M-1}{2M-1}\right)} \qquad (3.79)$$

$$\frac{\hat{v}(s)}{\hat{d}(s)} = \left[\frac{2V}{2M-1}\right]\left[\sqrt{\frac{M-1}{2\tau_L M}}\right]\left[\frac{1}{1 + sRC\left(\dfrac{M-1}{2M-1}\right)}\right]. \qquad (3.80)$$

By inspection, we see that both of these transfer functions have a single-pole characteristic, which will move as input voltage changes, thus altering the value of M. Also note that the right half-plane zero, which was present in the continuous mode, is *not* present in the discontinuous mode model for the boost SPC.

4. DC Transformers

The buck and boost SPCs are fundamental building blocks from which more complex power processing systems can be constructed. However, both of these fundamental SPCs have practical conversion limitations that can only be overcome by the use of another SPC element—*the DC transformer*. Recall that each of the basic converters can only accommodate one input and provide only one output with input and output sharing a common return reference point. Definite limits on their output voltage ranges relative to their input voltage values are evident. All of these limitations can be overcome by the use of additional DC transformation functions, the ideal form of which is depicted in Fig. 4.1. An ideal transformer element would pass all signal frequencies from DC to light with no power loss, provide any selected transformation ratio of voltage or current, provide complete isolation between its input and its output, process power in either direction with equal facility and, finally, provide for additional inputs or outputs if required.

Obviously, such a perfect transformation device does not physically exist and, fortunately, perfection is rarely required. Many circuits which approximate the characteristics of an ideal transformation element do exist and are quite practical, especially if unlimited bandwidth of signal processing is not required.

4.1. TYPICAL CONFIGURATIONS

One of the most common SPC circuits used as a DC transformer element in a power processor is shown in Fig. 4.2A. This circuit topology is often called a "push-pull" converter, or *parallel DC transformer*, in reference to the alternating conduction actions of the two primary-side switches (S1 and S2) and the secondary-side rectifier elements (D1 and D2). In Fig. 4.2A, the secondary rectifier connection is commonly called a *full-wave center-tapped* connection. However, there are other rectifier connections, such as the one shown in Fig. 4.2B. Here, this particular form is in a *full-bridge* connection. Regardless of their output rectifier connection, both circuits of Fig. 4.2 are classified as parallel DC transformers.

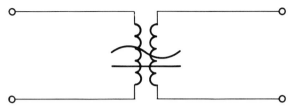

Fig. 4.1. The ideal transformer.

Switches S1 and S2 operate alternately, with conduction duty cycles of 50% during one complete switching cycle. These switching actions generate a symmetrical and alternating voltage across the primary of T1. Transformer T1 is a normal high-frequency AC transformer designed to operate at the switching rates of S1 and S2. Due to the AC voltage applied to N_p on each switching half

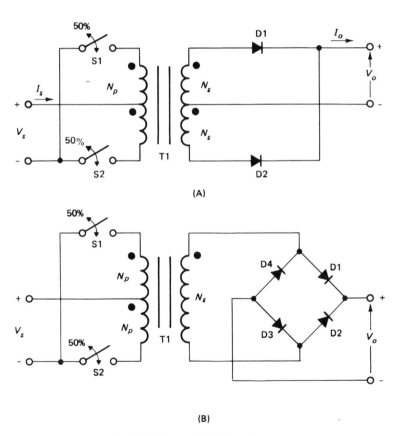

Fig. 4.2. The parallel DC transformer.

cycle, an AC voltage will appear on N_s such that:

$$V_o = V_s \left[\frac{N_s}{N_p} \right] \tag{4.1}$$

Transformer T1 provides desired voltage transformation ratio as well as input-to-output ground return isolation, if so desired. The corresponding AC voltages across the secondary windings of T1 are rectified by D1 and D2 so as to produce a DC voltage determined by Eq. (4.1). The circuits of Fig. 4.2 thus perform two series conversion steps, (DC-to-AC and AC-to-DC, respectively) to realize a DC-DC transformation function.

The voltage which appears across a nonconducting primary switch in this DC-DC transformer system is $2V_s$, as the voltage across one half of the total primary winding of T1 (N_p) must be reflected into its other half. The peak current in each of the two primary switches is equal to the input average current, I_s, when each is ON.

In addition to this parallel DC transformer circuit, three other primary switch variations are in general use today. These forms are commonly called *half-bridge, full-bridge*, and *single-ended* DC transformers and are shown in Figs. 4.3A through 4.3C, respectively, less associated secondary networks. There are many other primary switch connections possible for DC transformation, but the four variations just discussed represent the majority of schemes normally encountered in practice. More complex variations will be discussed in Chapter 9.

In the half-bridge circuit of Fig. 4.3A, switches S1 and S2 conduct alternately, i.e., when S1 is ON, S2 is OFF, and vice versa. The voltage appearing across any open switch is equal to the source voltage, and the peak current through any conducting switch is twice the average source current. Because of the reduced switch OFF voltage stress, this half-bridge primary connection is usually preferred over the parallel connection of Fig. 4.2A in offline applications since in the latter case, any open switch must ideally sustain twice the source voltage. On the other hand, for low source voltage values, the parallel connection is usually chosen, since any conducting primary switch sees only I_s rather than $2I_s$ in the case of a half-bridge DC transformer.

Notice that, in the half-bridge primary connection, the voltage across the primary of the transformer is ideally one-half the value of V_s. Therefore, for a given N_p, V_s, and V_o, the primary winding must have half as many turns.

In the full-bridge circuit of Fig. 4.3B, switches S1 and S4 are closed simultaneously during the first half-cycle of the conversion process, and then S2 and S3 are closed simultaneously during the second half-cycle. The voltage appearing across any open switch is equal to the source voltage and the peak current through any conducting switches is equal to the average source current. Because the full-bridge DC converter provides minimum switch voltage as well as minimum current stress, this approach is popular in high-power (>750 W) converters.

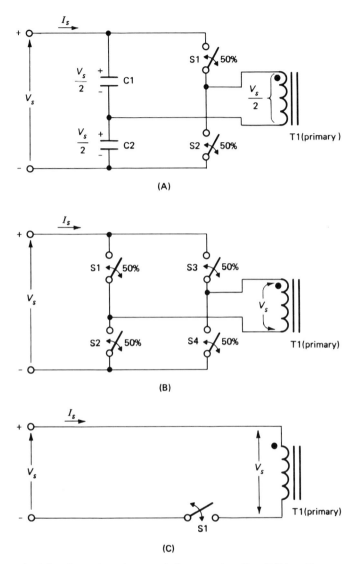

Fig. 4.3. Alternative primary switch connections for a DC transformer.

The single-ended approach to DC transformation (Fig. 4.3C) is simpler than all the others discussed, and is a very popular technique for low and medium output power applications. The circuit does, however, have some drawbacks. Its input current is pulsating, with an amplitude determined by the average value of I_s and the duty cycle of S1. Consequently, some form of input low-pass filter is usually needed to smooth out these current pulsations. When S1 is not conducting, its open-circuit voltage is equal to the source voltage plus the voltage reflected through the transformer from its secondary. In many cases, the switch open-circuit voltage will be greater than double the source voltage, while its closed-circuit peak current may well be more than twice the average input current. For these reasons, the single-ended DC transformer circuit is usually restricted to applications below 500 watts.

Finally, some means must be provided to reset the core of the transformer in this circuit when S1 is in an OFF state. This reset means is needed to prevent core saturation and subsequent high current in the switch when it is turned ON. Many transformer core reset schemes exist and may be divided into two general categories—those which allow the core energy to dissipate naturally, and those which force the reset of the core by external means. Which reset category is the best to use for a particular application is dependent on the core B-H loop characteristics, two extremes of which are shown in Fig. 4.4. The core material associated with the loop in Fig. 4.4A is characterized by a low remnant flux density (B_r) when its magnetizing force (H) is zero. When the magnetizing force is removed, this type of core can have a relatively large amount of energy stored. Here, an external means of core reset is desirable to quickly dissipate the large energy stored. The core characteristic of Fig. 4.4A is typical of some ferrite or powdered iron materials, or *any* core with an air gap. The characteristic shown in Fig. 4.4B is typical of cores composed of grain-oriented nickel-iron alloys, with no air gaps. Because of its "square-loop" B-H characteristic, this type of core material stores little energy and has a high B_r at zero excitation.

Comparing the two B-H characteristics, we see that each has advantages and disadvantages. With a low B_r core, the reset scheme can be very simple which is a significant advantage. The disadvantages of a low B_r core include the need for substantial core area and/or winding turns to keep the magnetizing current within reasonable bounds and the fact that less than half the available core flux can be used. A high B_r core, on the other hand, may require a more complex reset arrangement but, in exchange, will have a smaller core and/or less copper in the windings. As a general rule for the same power level, the forced reset high B_r core will provide the smallest complete transformer.

In practice, gapped low-B_r cores are most frequently used in single-ended DC transformer converters. The reason for this choice is not necessarily that this core characteristic is superior for this application, but rather due to a general lack of appreciation by designers of how simply forced-current reset can be implemented as well as the advantages of so doing.

A wide variety of reset schemes are possible, with simple clamp methods as shown in Fig. 4.5. In theory, the diode-zener clamp may be added to any winding of the transformer to provide core reset, provided that the associated volt-seconds/turns factor is sufficiently large that the core is indeed reset during the *minimum* time period that the switch is OFF. In practice, however, the clamp is usually placed on the primary winding (Fig. 4.5A), since any practical transformer

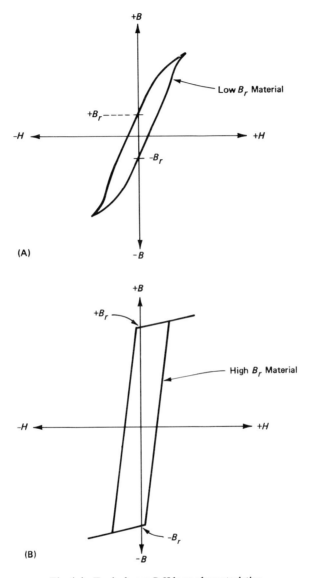

Fig. 4.4. Typical core B-H loop characteristics.

will have some primary-to-secondary leakage inductance. This parasitic inductance will also store energy and can damage the switch element if not dissipated
safely. By placing the clamp circuit across the primary winding, both core reset
and switch protection against parasitic transformer leakage inductance effects
are accomplished simultaneously. Of course, in this type of core reset circuit,

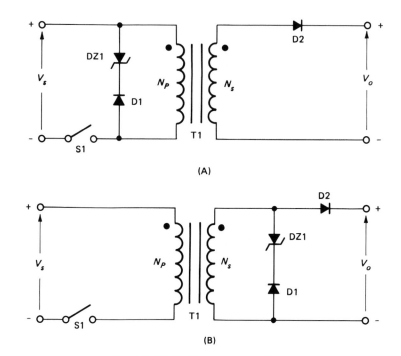

Fig. 4.5. Dissipative core reset schemes.

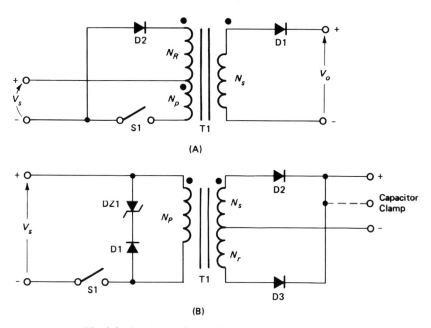

Fig. 4.6. Core reset schemes that recover the core energy.

all of the stored core energy is lost within the clamp network. For this reason, this simple reset scheme is suitable for low-power applications.

For higher power applications, the stored energy of the transformer's core may be either returned to the source voltage (Fig. 4.6A) or to the load (Fig. 4.6B). In this manner, only a small portion of the stored core energy is lost, and the input-to-output power conversion efficiency is high. There is no requirement that N_R be equal to N_P or N_s. In fact, by making N_R small compared to N_P, the duty cycle of S1 can be increased with a reduction in the peak current it must handle. Unfortunately, as N_R is made smaller, the open-circuit voltage (V_a) across S1 will increase. The expression for V_a versus V_s is:

$$V_a = V_s \left[1 + \frac{N_P}{N_R} \right]. \tag{4.2}$$

In practice, the maximum switch duty cycle (D) of the single-ended transformer is usually limited to 0.7 or 0.8. For example, if the reset scheme of Fig. 4.6A is used with a maximum duty of 0.8, the number of turns on the primary must be more than *four* times the number of turns on the reset winding. From Eq. (4.2), the ideal switch voltage stress when it is in an OFF state will then be at least *five* times the source voltage.

In Fig. 4.6B, notice that a dissipative clamp circuit is shown on the primary of the transformer. This voltage clamp is needed to protect S1 from the harmful effects of parasitic primary leakage inductance and does not play a major role in the core reset operation. Power dissipation in this clamp circuit is normally quite small. When the core energy is discharged into the secondary as shown in Fig. 4.6B via diode D3 it will be necessary to connect D3 to a low-impedance load, such as the capacitor of any associated low-pass filter network. Connecting the energy commutation diode to a high-impedance load (such as an inductor) will cause a large voltage spike to appear on both primary and secondary windings of the transformer.

When a core of a single-ended DC transformer has a high value of B_r, mechanisms can be provided for reset. Two possibilities are shown in Fig. 4.7, the first of which uses an auxiliary winding and a current source and the second a permanent magnet. Ferrite cores having an integral permanent magnet are now available, but are rarely employed (primarily because of cost of customization) despite the simplicity of the reset scheme. The current reset scheme of Fig. 4.7A is not often used in practice because of the need for an additional transformer winding and an external current source. However, in many SPC circuits, there is a built-in current source—the filter inductor of any associated output low-pass network! In these instances, current reset may be realized by simply tapping the secondary of the transformer, as shown in Fig. 4.8, and routing the inductor current through the tap via a diode. This method costs little to implement—a tap on the transformer. In most circuits the diode will already be present.

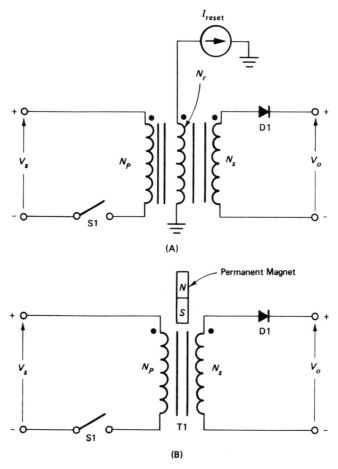

Fig. 4.7. Forced core reset schemes.

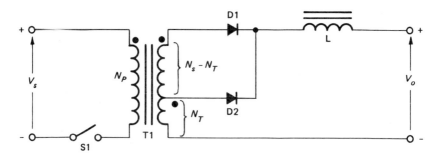

Fig. 4.8. Current reset using an output inductor.

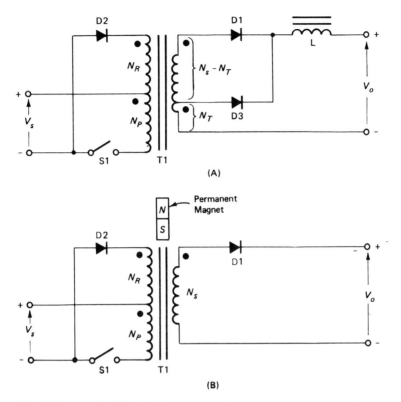

Fig. 4.9. Practical realizations for current and permanent magnet reset schemes.

The reset schemes of Figs. 4.7 and 4.8 lack one important feature, namely, a means for limiting the amplitude of the OFF voltage stress on S1 (V_a) during core reset. This feature can be easily added by the use of the primary clamping scheme of Fig. 4.6A. The resultant circuit structures are shown in Fig. 4.9.

In converter applications which have a high-voltage source, V_a can become substantial, as Eq. (4.2) predicts. A means to minimize V_a in these applications is shown in Fig. 4.10A. In this scheme, two primary switches are used and are driven ON and OFF simultaneously. Here, V_a for each switch will be equal to V_s, and the maximum duty cycle is 0.5, since the primary winding is also the reset winding. The duty cycle limit may be increased, again at the expense of increasing V_a, by tapping the primary winding as shown in Fig. 4.10B.

4.2 UNIDIRECTIONAL DC TRANSFORMERS

As we have just seen, there are a wide variety of circuits that can approximate the properties of an ideal DC transformer element. Here, we will look at SPC DC

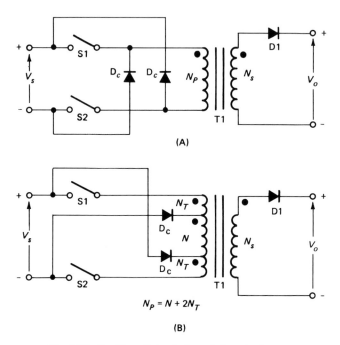

Fig. 4.10. Double-switch, single-ended reset schemes.

transformer circuits in which the input-to-output power flow is *unidirectional*. Then, in the next section, similar circuits in which *bidirectional* power flow is possible will be examined.

When comparing the DC transformer circuits to follow, one must take into consideration more than just topology concepts. As a practical matter, both the choice of switch arrangements and the means for driving them must be considered. There are several choices to be made in switch-drive arrangements alone, i.e., external drive or self-oscillating, voltage or current-drive, and many more.

One of the simplest (and oldest), unidirectional DC transformers is the *Royer* circuit shown in Fig. 4.11A. This is a self-oscillating BJT circuit, in which the transistors are driven from positive-feedback auxiliary windings on T1. For the circuit to be self-oscillating, it is necessary that the transformer core saturate at the end of each half cycle, removing the base drive from the switch in conduction. Residual energy in the transformer's core changes the polarity of the voltage across its transformer windings when a switch turns OFF, turning on the opposing switch. The alternate switch remains in conduction until the core saturates once again, and the process repeats. The primary advantage of the Royer circuit is its simplicity, but it does have a number of practical drawbacks which

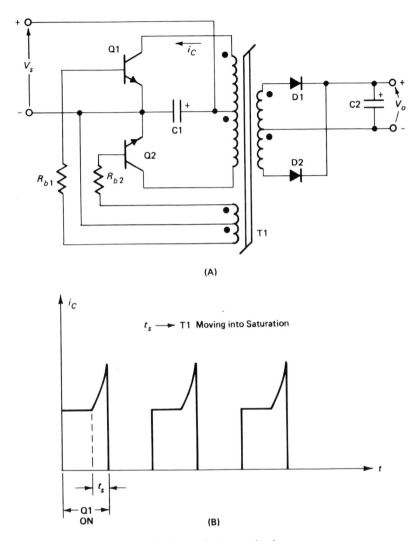

(A)

(B)

Fig. 4.11. The Royer circuit.

limits its use to low-power applications. One major problem occurs when the core goes into saturation reducing the primary impedance to a low value. This causes a large increase in i_C during the interval when the switch is turning OFF, as shown in Fig. 4.11B. This current transient causes high peak power to be dissipated in the transistor. For this reason, the Royer circuit is preferred for low-power DC transformer applications where such peak transient currents are small.

The oscillation frequency of the Royer approach is more or less a linear function of the source voltage. It will also change as a function of temperature since the core flux saturation level, V_{CE} (sat) and the winding resistances of the transformer are all directly related to the environment. An additional problem is that the core flux excursion must go from $+\phi_s$ to $-\phi_s$, thus limiting the maximum self-oscillating frequency due to core power dissipation.

Another topology similar to the Royer circuit, which removes the switch drive functions from the main power transformer, is shown in Fig. 4.12A. This circuit is called the self-oscillating *Jensen* converter. In this approach, converter oscillation frequency and base-drive functions are performed by the saturable

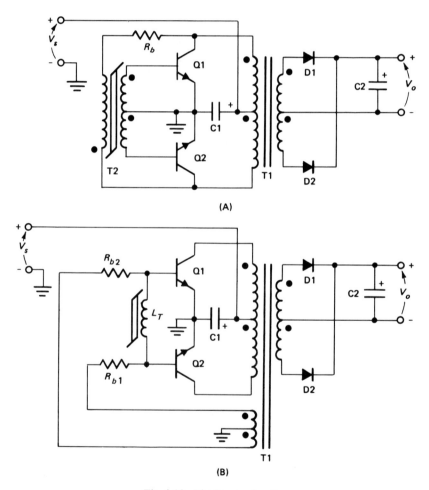

Fig. 4.12. The Jensen circuit.

transformer element, T2. Therefore, T1 can be optimized for power tranfer alone. The oscillating frequency of the Jensen approach is determined by the core characteristics of T2 and the base-emitter voltage of the transistor switches. These voltages will vary with both converter load and ambient temperature, but normally not enough to have a large effect. The oscillation frequency of the Jensen converter is relatively stable with line, load, and temperature changes when compared to similar conditions in a Royer circuit. The primary switching current waveforms of the Jensen converter are also more uniform, with little or no current spiking due to core saturation. The price paid for this improvement is the need for an additional transformer. The Jensen circuit can be simplified as shown in Fig. 4.12B, where the timing function is accomplished by a single-winding saturating reactor (L_T), with the base drive power for the switches provided by an auxiliary winding on the power transformer, T1.

If the load and/or source voltage of the Jensen converter varies over a wide range of values, the transistor base drive approaches of Fig. 4.12 may be unsuitable, since they provide for drive magnitudes which are proportional to the source voltage. To provide sufficient base drive under worst-case input/output line and load conditions, (minimum V_s, with maximum load current), the base resistance (R_B) must be selected to provide enough current to maintain the switch in a fully saturated condition. Thus, when V_s is at its maximum and the converter load current is at a minimum, the switches are grossly overdriven. Long transistor storage times result. To maintain a base drive proportional to load demand, a current-drive network must be employed, as shown in Fig. 4.13A. In this circuit:

$$i_b = i_C \left[\frac{N_2}{N_1} \right] - I_{m2} \tag{4.3}$$

where I_{m2} is the magnetizing current in winding N2 of T2. By using a proportional base current drive such as this, the switches are always operating at a fixed β except when the converter has low output currents. Under light load conditions, i_C is small, and I_{m2} may become large enough to make the resultant i_b small. A low i_b value could result in inadequate base drive to maintain transistor saturation, and produce high power dissipation in these semiconductors. To overcome the light-load drive problem, an additional winding may be added to T2 to compensate for I_{m2}, as shown in Fig. 4.13B.

Besides providing excellent switching characteristics, the proportional base drive network of Fig. 4.13 is very efficient. For these two reasons, these forms of self-oscillating parallel DC transformers can work well at power levels up to several hundred watts. However, proportional current drive does have a disadvantage to keep in mind. If the output of a Royer converter is short-circuited, the circuit will cease oscillation and remain quiescent with both of its primary switches OFF. To a large degree, the Royer circuit is self-protecting in this

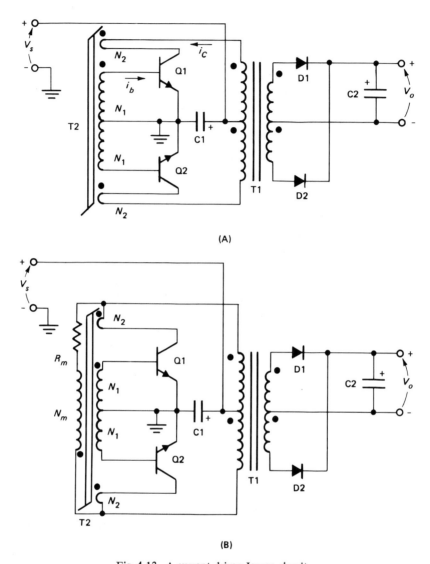

(A)

(B)

Fig. 4.13. A current-driven Jensen circuit.

regard. Similarly, the voltage-driven Jensen converters of Fig. 4.13 are also somewhat self-protecting in overload situations. This is not to imply that for all conceivable conditions of output current overload that the circuits of Figs. 4.11 and 4.12 will always be self-protecting by nature but, in general, the degree of self-protection is very good. However, in the current-driven circuits of Fig. 4.13, the feature of self-protection from output overloads is notably absent except, perhaps, for a complete short circuit condition on their outputs. As the output load increases in value, I_b increases proportionately in these schemes. Therefore, the proportionality feature of current drive will then cause the collector currents of the switches to reach peak stress levels, usually resulting in damage, unless the transistors are turned OFF by external protective means.

When efficient high power processing at a controlled frequency of conversion is desired, an externally driven DC transformer is usually used. For example, the circuit in Fig. 4.13A could be externally controlled by simply adding another center-tapped winding to T2 and connecting two drive switches as shown in Fig. 4.14. At the beginning of the switching cycle, S3 and S4 are both ON, disabling the transformer action of T2. This prevents S1 and S2 from turning ON. Driver switch S3 is then turned OFF and the residual energy stored in T2 (by I_D) is discharged into the base of primary switch S2 turning it ON. S2 conduction will then be maintained by the proportional base current supplied by the positive current feedback winding (N_2) of T2. At the end of this half cycle, S3 again turns ON, shorting the drive winding of T2, and disabling the proportional base drive mechanism for S2. One switching cycle is completed when similar actions are accomplished with S1 and S4. Although we are concerned here with 50% conduction cycles, the conduction periods of S1 and S2 could be pulse-width-modulated (PWM) if a low-pass LC filter is provided on the output of this converter for energy storage when switches S1 *and* S2 are both OFF.

Both full-bridge and half-bridge externally-driven DC transformer variations are frequently used in practice. An example of a half-bridge version is shown in Fig. 4.15. Here, only a single positive feedback winding on T2 is needed to accommodate proportional base drive for both S1 and S2.

DC transformer drive schemes abound in practice. Unfortunately, we cannot present them all in this text. The ones covered do represent the major schemes

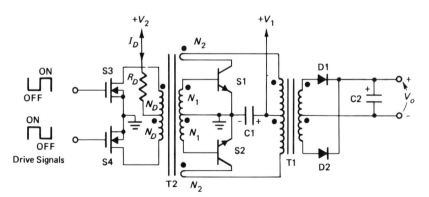

Fig. 4.14. A driven parallel circuit with proportional base drive.

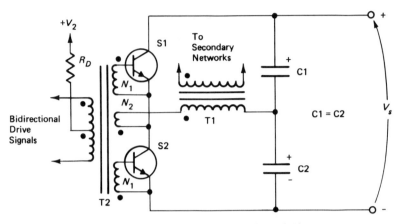

Fig. 4.15. A half-bridge version of Fig. 4.14.

found in today's conversion systems. Which scheme will be the best for a particular application is a function of conversion needs, size, and, of course, cost.

4.3. BIDIRECTIONAL POWER FLOW

For some power conversion applications, such as battery charging and discharging, it is desirable for the DC transformer to have bidirectional power flow capability. These features can readily be achieved by making appropriate converter switch elements bidirectional with corresponding drives for directional states. An example of adding this feature to the parallel transformer of Fig. 4.2 is shown in Fig. 4.16. Here, we have chosen MOSFETS as switch elements, as they possess diodes as an integral part of their structure. The direction of power flow will depend on the values of V_s, V_o, N_P, and N_s, as indicated by the following relationships:

$$V_s > \left[\frac{N_P}{N_s}\right] V_o \Rightarrow \text{power flow from input to output} \tag{4.4}$$

$$V_s < \left[\frac{N_P}{N_s}\right] V_o \Rightarrow \text{power flow from output to input.} \tag{4.5}$$

By making the switches bidirectional in handling currents, the discontinuous mode of operation is *not* possible. As the inductor current approaches zero during a switching cycle, it will continue on, becoming *negative* rather than remaining at a zero value. This means that at $t = T_s$, $I_a \neq 0$ but may have some negative value. Therefore, the continuous mode of SPC operation and small-

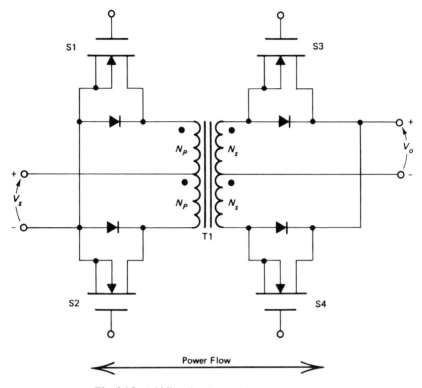

Fig. 4.16. A bidirectional parallel DC transformer.

signal modeling associated with it discussed in Chapters 1 and 2 will *always* prevail in a converter with bidirectional power flow capability.

4.4 PROBLEMS AND LIMITATIONS OF PRACTICAL DC TRANSFORMERS

Obviously, practical DC transformers can not exhibit ideal electrical character-istics and, as such, have performance limitations and operational problems. For-tunately, most of the circuit problems can be overcome and the functional limita-tions minimized.

The most fundamental performance limit is bandwidth of signal processing. For practical purposes, the bandwidth can be assumed to be less than one-half of the maximum switching frequency of the converter. It follows that, to achieve wide bandwidth of power processing, high switching frequencies must be used. At the present time, conversion switching frequencies up to 500 kilohertz are not uncommon and, by using MOSFET switch elements, frequencies of several megahertz are possible. For most applications, converter bandwidth is not a problem if associated active and passive circuit elements are properly chosen or designed.

Any practical switch element requires some finite period of time to change state. In the BJT, the delay in turnoff is primarily due to minority carrier storage time. One example of the effect of BJT storage time is conduction time overlap in the switches of a DC transformer. The result of finite switching times and conduction overlap in a parallel DC transformer (Fig. 4.17B) is shown in Fig. 4.17A, where its output secondary waveform appears much like that shown somewhat exaggerated in Fig. 4.17A. Two design approaches can be followed to reduce the effects of these "breaks" in output voltage. First, reduce the con-

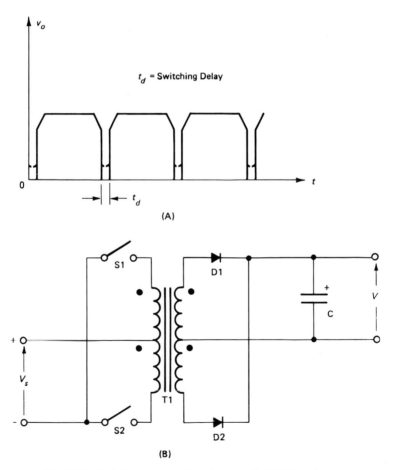

Fig. 4.17. Typical output waveform in a practical DC transformer.

duction overlap to acceptable intervals and, second, add a small capacitor (Fig. 4.17B) across the output terminals of the circuit to smooth out the waveform pulsations.

Conduction overlap of primary switches also causes another problem discussed earlier, namely, high current spikes during such intervals. If the transformer is supplied from a "stiff" voltage source and both switches are ON simultaneously, there is very little impedance to limit the switch current amplitudes and they could very well be damaged or destroyed as a result. If converter drives are externally provided, their signal intervals may be shortened to provide a *dead time* when neither primary converter switch is being actively driven. The amount of dead time required will be determined by maximum storage time of the primary switches. Because the storage time of a BJT is a function of base drive magnitude, collector current level, junction temperature and normal variations in the manufacturing processes of the semiconductor, the required dead time intervals are difficult to predict and, to satisfy the worst case, may have to be a significant portion of the switching cycle of the converter. Any output filter capacitor for energy storage (Fig. 4.17A) also will end up being fairly large.

In a self-oscillating DC transformer circuit, automatically inserting an adequate dead time interval to prevent conduction overlap of primary switches is often unrealistic. Usually, some form of in-line primary current-limiting is added instead to reduce the amplitudes of any potentially destructive current spikes to safe levels. A simple means for limiting instantaneous primary current is the insertion of an inductor in series with the source voltage as shown in Fig. 4.18. In Fig. 4.18A, the energy stored in L following any overlap interval of S1 and S2 is dissipated in resistor R_o and diode D_C. In Fig. 4.18B, the stored energy in L is returned to the source via diode D_C.

Large current spikes in the primary switches can also occur should the transformer core saturate when either one is conducting. In an ideal converter, the volt-seconds applied to both the primary and the secondary windings will average to zero over one switching cycle and the corresponding flux excursions of the transformer core will be symmetrical about the origin of its hysteresis loop. Therefore, no core saturation problems will occur as long as maximum flux capability of the core is not exceeded. In practical terms, however, perfect volt-second balance is never achieved for a number of reasons. First of all, the conduction interval of the primary switches may not be exactly identical and their conduction voltage drops may not be equal. Also, conduction voltage drops as well as conduction time intervals of rectifier diodes on the secondary side of the transformer may not be equal because of parasitic inductances. This problem will be addressed in more detail in Chapter 8.

There are many ways to minimize or eliminate transformer core saturation from unbalance and its resultant undesirable side effects. Some techniques are simple in concept while others can be very complex to implement. The simplest

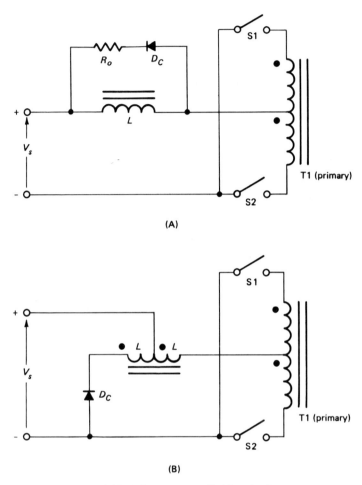

Fig. 4.18. Primary current-limiting circuits.

method is the selection of a core material for the transformer that has relatively low permeability and low residual flux density (such as is shown in Fig. 4.4A) which will tolerate some DC offset without saturation. This solution brings with it relatively high magnetizing current levels for small core area dimensions or low numbers of turns on windings. Increasing core area and winding turns to reduce magnetizing current levels increases the size and cost of the transformer. A better alternative would be the use of a core with a characteristic like that shown in Fig. 4.19A. In this type of core characteristic, the effective permeability is very high in the normal operating region (about the origin of the B-H loop). If some DC imbalance is present and the operating loop moves off to one side of the core

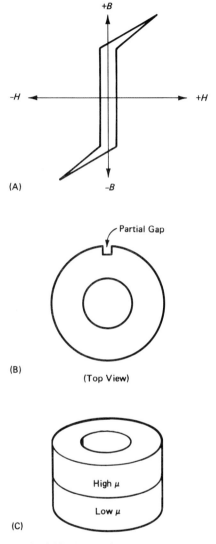

Fig. 4.19. Composite core structures.

characteristic, the core does not go into hard saturation, but moves into the upper region of the characteristic, where the permeability is lower. Thus, sufficient inductance will exist to limit peak switch currents to safe levels. While there is no known single core material with the desirable characteristic of Fig. 4.19A, modified or composite core structures can be built which can emulate it. Two examples, using toroidal structures, are illustrated in Figs. 4.19B and C. In

Fig. 4.19B, a partial air gap is placed on the outer part of the core so that nearly all of the core flux is contained in its inner ungapped section during normal operation. If the inner section should saturate, the core flux will move to the outer air-gapped portion which has a lower but still significant permeability. A stacked two-core composite configuration is shown in Fig. 4.19C, where two cores with the same inner and outer diameters but composed of different permeability materials are used. The methods of primary current-limiting (shown in Fig. 4.18) for overlapping conduction will also provide switch protection for core saturation.

For those DC transformation circuits which use a bridge or half-bridge primary switch connection, a particularly simple means is available for eliminating transformer core saturation due to primary volt-second imbalance. In these two converter circuits, the primary winding of their transformers can be DC-isolated by the insertion of a series capacitor (C_s) as shown in Fig. 4.20. Notice that for the half-bridge version (Fig. 4.20B), the series DC-blocking capacitor function is accomplished inherently by the presence of the voltage-division capacitors, C1 and C2. The value chosen for C_s cannot be arbitrary, for as C_s is made smaller, the output voltage waveform (Fig. 4.20C) will become more tilted because of the AC voltage drop across this series capacitor. On the other hand, if C_s is made too large and the source voltage is subject to rapid changes, the DC voltage across C_s may not change quickly enough to prevent momentary transformer core saturation. The time constant (τ_C) for the rate of change of DC voltage across C_s is approximately:

$$\tau_C \cong R_r C_s \qquad (4.6)$$

where R_r is the reflected output load resistance of the converter seen by the primary of the transformer.

If possible, the value of τ_C should be made small in comparison to the maximum slew rate of expected changes in V_s.

A variety of active circuits are possible that will electronically sense the condition of impending core saturation and turn off primary switch networks. The price paid for any electronic approach is additional SPC complexity as well as cost.

The possibility of transformer core saturation can be reduced if converter switches are closely matched in thermal as well as electrical characteristics. Matching can greatly improve the protective effectiveness of the passive methods discussed earlier. For MOSFET switches, matching is not difficult or unreasonable from a cost standpoint. However, matching BJTs requires that several parameters (V_{CE} (sat), V_{BE}, t_s, etc.) be controlled over the ranges of junction temperatures and load currents encountered in an application. BJT matching is

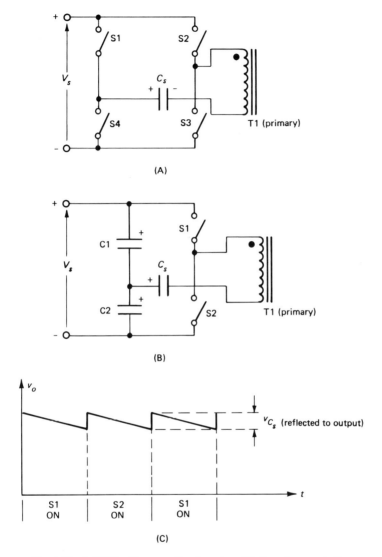

Fig. 4.20. A DC-blocking capacitor in series with the primary winding.

rarely economically practical, although pair matching for some parameters will always be to the designer's advantage.

When a DC transformer circuit is shut down by rapidly removing its source of power or its drive, the transformer core may not have zero residual flux but instead be left with a large value of residual flux. When the converter is reactivated, the core material of the transformer may saturate on the first half-cycle of operation, resulting in a potentially destructive current spike in one of the primary switches. Since there is no means in most SPC circuits to determine the actual initial value of the core flux at converter turn-on, this particular problem has destroyed more than a few converters. Both inductive primary current-limiting methods or active core state flux-sensing schemes are effective approaches in solving this turn-on core saturation dilemma.

The self-oscillating transformer circuits shown in Figs. 4.11 through 4.13 work very well, once oscillation has begun. However, all of these circuits suffer from a lack of reliable "starting." Since both primary switches are initially OFF, there is no inherent mechanism (other than leakage paths) to start oscillation at power application. Therefore, it is necessary to include auxiliary starting electronics which momentarily place one (or more) of the DC transformer switches in an active conduction mode. A commonly used starting circuit for the Royer system of Fig. 4.11 is shown in Fig. 4.21. Many other starting networks exist. Interested readers are referred to text reference 5 in the bibliography of this book.

In the parallel circuit of Fig. 4.2, if the two primary windings of the transformer are not tightly coupled, it is possible to generate switch voltage spikes from leakage inductances. The effect is equivalent to placing a small discrete

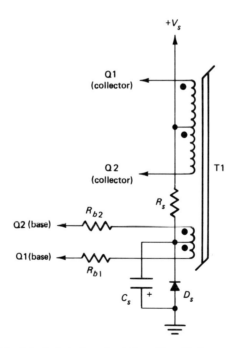

Fig. 4.21. A typical starting circuit for self-oscillating converters.

inductor (L_ϱ in Fig. 4.22) in series with each primary switch. When a switch is turned OFF, the energy in the associated parasitic leakage inductance must be dissipated. For this reason the voltage across the OFF switch will continue to rise until this energy is eliminated, usually as a result of switch breakdown. If the transformer leakage inductances are made very small, then any parasitic capacitance in parallel with the OFF switch will limit the resultant transient voltage amplitude, often protecting the switch from voltage breakdown and producing a damped ringing waveform which can contribute to the electro-magnetic (EMI) noise generated by the converter. Winding the primary windings of the transformer either bifilar or interleaved is a common design method used to maximize coupling between the windings. Adding a transient-damping net-work (R_s, C_s) across the primary winding, as illustrated in Fig. 4.22, is also recommended.

When the output load of a practical DC transformer is made very light, voltage

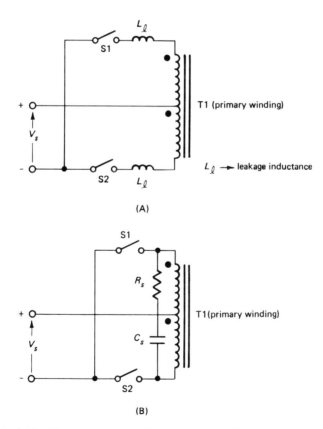

(A)

(B)

Fig. 4.22. Primary-to-primary leakage inductance with a damping network.

transients resulting from winding and circuit layout parasitic inductances can appear as a dynamic part of the output voltage, altering its average voltage value, independent of the value of V_s. This transient-produced output problem is best resolved by careful design of the transformer, good physical layout of converter power electronics and appropriate damping networks across parasitic inductive elements.

The magnetizing inductance of a practical transformer of a converter must be finite, and therefore contributes to the primary current waveform. A general representation of the AC waveform of the magnetizing current waveshape is shown in Fig. 4.23A. Since i_m is an inductive current, it does not represent an inherent conversion power loss. However, from earlier discussions, it can increase the converter switching losses and the resistive losses in the winding of the transformer. At full load, I_m is normally small in amplitude when compared to I_s (as shown in Fig. 4.23B), so that their composite waveform is always positive. However, at light output loads, I_m may exceed I_s and cause the composite current waveform to reverse polarity. A typical solution to this light-load problem is to make the primary switches bidirectional. For BJT switches, anti-parallel diodes can be added as shown in Fig. 4.24. During light loading when the diodes are conducting, the effective duty cycle of the switches may be different from that applied to the switch element alone. If the switch is PWM controlled, then the voltage transfer function will be altered.

In a practical single-output circuit, each half section of the transformer secondary winding is not perfectly coupled to either the primary winding or to its other half section. In addition, physical connections to output rectifiers will always have some series inductance associated with them. These parasitics can affect the output voltage regulation and ripple of the converter. The equivalent circuit of an SPC output secondary winding with rectifier connections can be drawn with these parasitics, with leakage inductances drawn as lumped inductors in series with each rectifier. The resultant circuitry is shown in Fig. 4.25A. Note that there are two types of inductances illustrated—that which is common to both halves of the secondary winding and that which is due to imperfect transformer coupling between these secondary windings plus their rectifier connection inductance (L_a and L_b). For the purposes of this discussion, the presence of the common inductances will be ignored and the equivalent circuit of Fig. 4.25B used. These parasitic inductances do affect converter circuit operation, becoming more significant as either the switching frequency or the output current level is increased.

The voltage and current waveforms within the equivalent circuit of Fig. 4.25B are given in Fig. 4.26, for the case of 50% duty cycle on each secondary. During time t_1, a current i_1 is flowing through D1 and L_a. At $t = t_1$, the polarity of v_1 and v_2 relative to the centertap of the transformer secondary changes, but, because of the presence of inductors L_a and L_b, i_1 will fall gradually and i_2 will increase gradually. The average value relationship,

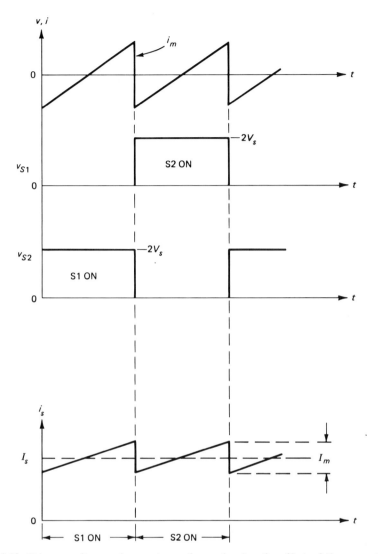

Fig. 4.23. Primary voltage and current waveforms showing the effect of the magnetizing current.

$$I_1 + I_2 = I_3 \qquad (4.7)$$

must be preserved throughout the converter's operating cycle. During the interval Δt, when both i_1 and i_2 are flowing, D1 and D2 must be in conduction simultaneously, so that $v_3 = 0$. This "notch" in v_3 reduces the average output

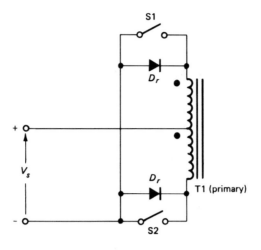

S1, S2 $\longrightarrow$ undirectional switch elements

Fig. 4.24. Primary free-wheeling diodes.

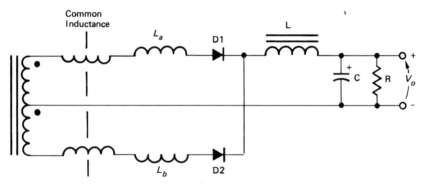

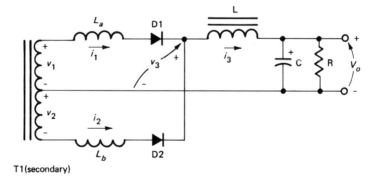

Fig. 4.25. Secondary equivalent circuits.

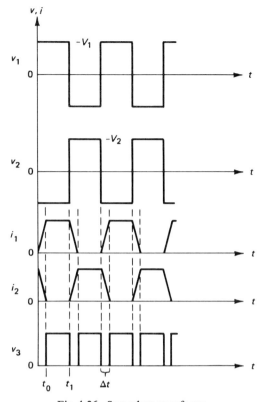

Fig. 4.26. Secondary waveforms.

voltage of the converter and increases associated AC ripple magnitudes, both of which are undesirable.

The overlap time interval (Δt) is:

$$\Delta t = \frac{L_a i_1}{V_1}.$$ (4.8)

It is obvious from Eq. (4.8) that low-voltage high-current transformer secondary windings are most likely to produce a significant value of Δt. The "delay" interval will also vary with the converter's load current. This time interval can be related to a change of conversion duty cycle, ΔD, and ripple frequency ($f_r = 2f_s = 2/T_s$) by:

$$\frac{\Delta t}{T_s} = \Delta D = \frac{L_a i_1 f_r}{2V_1}.$$ (4.9)

For conversion duty cycles less than 50%, the "notch" effect is still evident, even though v_1 goes to zero prior to change in secondary voltage polarity. Therefore, i_1 must continue to flow to satisfy the requirements of Eq. (4.7). In practical circuits, i_2 may not remain zero during the Δt time interval and i_1 is thus reduced. This situation can reduce ΔD to some degree, but the basic result is still evident. Voltage v_3 will be zero during Δt only if $L_a = L_b$. Usually, this latter condition is not satisfied. Therefore, voltage v_3 usually has some positive value during Δt. The net result of all of the above actions is a reduction in effective duty cycle of the v_3 waveform when compared to that of the transformer secondary voltages. The change is dependent on load current.

In a DC-DC converter with multiple secondary outputs, transformer cross-regulation between secondaries is degraded by this effect as well as by parasitic resistances of rectifier diodes, winding resistance voltage drops, and other leakage inductance effects. For good transformer winding cross-regulation, it is imperative that parasitic series inductances be minimized.

So far we have been dealing with symmetric output converter circuits with equal parasitic inductances, i.e., $L_a = L_b$. Now consider a single-ended DC transformer output circuit, such as that shown in Fig. 4.27A. Here, L_a will most likely be different in value than that of L_b. The resulting waveform shape of v_3 will be very similar to that shown in Fig. 4.27B. The corresponding equation for Δt_1 and Δt_2 are:

$$\Delta t_1 = (L_a + L_b)\frac{i_1}{V_1} \tag{4.10}$$

$$\Delta t_2 = (L_a + L_b)\frac{i_1}{V_1'} \tag{4.11}$$

and, because v_1 is usually asymmetrical, $\Delta t_1 \neq \Delta t_2$. The equations for v_3 during Δt_1 and Δt_2 are, respectively:

$$v_3 = \frac{L_b i_3}{\Delta t_1} \tag{4.12}$$

$$v_3 = \frac{L_b i_3}{\Delta t_2}. \tag{4.13}$$

4.5 SOURCE IMPEDANCE EFFECTS

So far in the discussions of this chapter, we have tacitly assumed that a voltage source was being used at the input to the DC transformer, as shown in Fig. 4.28A. This is referred to as a *voltage-fed* transformer. Often the power source

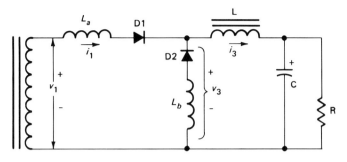

T1(secondary)

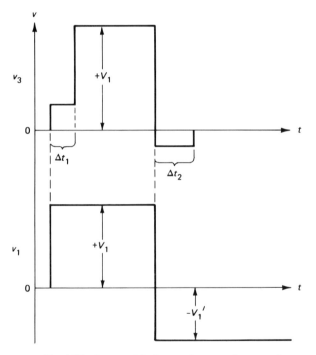

Fig. 4.27. Asymmetrical converter secondary waveforms.

is high in impedance value, and as such could be characterized as a current source, as shown in Fig. 4.28B. One practical and common form of a "current" power source is shown in Fig. 4.28C. In this instance, the transformer system is said to be *current-fed*.

In the case of voltage-fed DC converters, voltages appearing across the transformer windings are determined by source voltage and the winding currents by

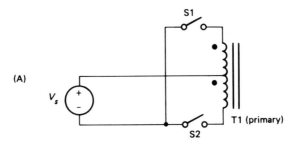

(A)

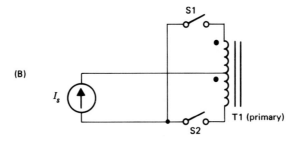

(B)

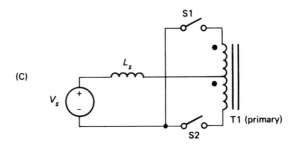

(C)

Fig. 4.28. Voltage- and current-fed transformers.

the converter load. In the case of current-fed converters, transformer currents are determined by the impedance of the source and winding voltages by the load of the converter. Obviously, a dual source relationship exists between these types of converter feed.

Taking into account the parasitic effects, such as switch conduction overlap and others discussed earlier in this chapter, a current-fed converter has many desirable features, since it automatically limits primary currents. Also, the current-fed converter is less prone to output capacitor charging under light output load conditions. On the other hand, the current-fed converter does have its disadvantages. For example, any practical transformer will have some leakage inductance between primary and secondary windings which, can produce large voltage spikes across converter switches when they are turning OFF. Again, good magnetics design and liberal use of appropriate transient damping networks are the best defenses. Also, when a self-oscillating converter is operated from a current-limited source, a starting problem can exist because of starting-voltage dependency on output load magnitude.

5. Buck-Derived Circuits

In Chapter 1, we asserted that complex SPC circuits could be reduced to equivalent forms of one or more combinations of the buck, boost, and the DC transformer element. Because of that assumption, Chapters 2, 3, and 4 were devoted to detailed examinations of the characteristics of these three basic converters. Now the time has come to test our original premise and to see if this evolutionary point-of-view is valid and useful in practical SPC design.

In this chapter, we begin by examining the family of SPC circuits which can be derived from the basic buck converter by the insertion of different DC transformers into its topology at various points. In the pursuit of these combinational investigations, opportunities to reduce the switch count will arise, whereby other SPC circuit changes will become apparent that, in turn, will reveal further offspring of the buck SPC family.

The first combinations we will examine are those which use the parallel DC transformer of Fig. 4.2A, followed by similar circuit variations using bridge, half-bridge, and single-ended DC transformers. We will digress a bit in the related discussion on the single-ended transformer variation to point out the consequences of using the different transformer core reset schemes of Chapter 4 on SPC circuit performance.

5.1 BUCK SPC AND PARALLEL TRANSFORMER COMBINATIONS

Figure 5.1 shows the two basic circuits under consideration—the buck converter (A) and the parallel DC transformer (B). In Fig. 5.1A, we have indicated five circuit locations (A-A' through E-E') where insertion of a DC transformer is feasible so as to be in series with the flow of input-to-output power. Other insertion areas within the buck SPC topology for DC transformers are conceivable but, in general, yield converter systems that are not functional, as the DC transformer would not be in series with the power flow path.

If we insert the DC transformer at location A-A', the converter shown in Fig. 5.2A is obtained. Here the buck converter usually serves as a regulated voltage source for the DC transformer. This system is usually referred to by

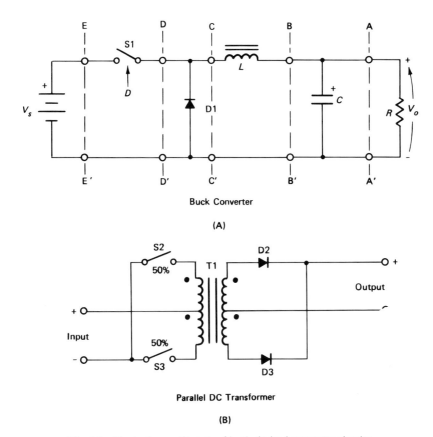

Buck Converter

(A)

Parallel DC Transformer

(B)

Fig. 5.1. The basic constituents of buck-derived converter circuits.

SPC designers as a *DC-DC converter with a pre-regulator* and can be altered to include multiple-output capability, as shown in Fig. 5.2B. Control of the conduction duty cycle of S1 is usually maintained by auxiliary voltage control electronics associated with either the output of the pre-regulator (the buck converter portion) or one of the outputs of the DC transformer. Keep in mind that the DC transformer is voltage-fed, and one may experience the unbalance and conduction overlap problems discussed in Chapter 4.

Even though the circuit of Fig. 5.2 uses a buck converter as a pre-regulator, the addition of the DC transformer allows the output voltage(s) to assume a value(s) higher or lower than the source voltage (V_s). DC isolation between input and output is now possible and, as shown in Fig. 5.2B, a number of different output voltages can be provided simultaneously. The output voltages are related to the output voltage of the pre-regulator buck SPC by the corresponding turns ratio of the transformer windings.

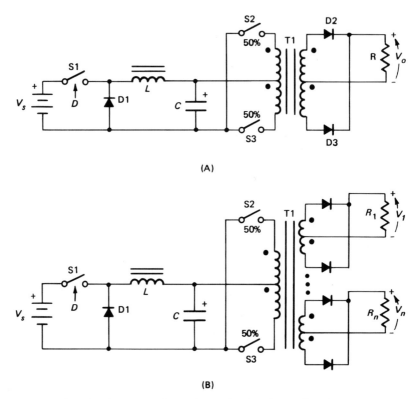

(A)

(B)

Fig. 5.2. The DC transformer inserted at location A-A' in Fig. 5.1.

Even though the range of voltage gain (M) is no longer restricted to less than one and input-to-output DC isolation has been achieved, the converter of Fig. 5.2A (and those to follow in this chapter) still retains the basic properties of a buck SPC.

When the DC transformer insertion point is moved to position B-B' of Fig. 5.1A, the converter shown in Fig. 5.3A results. This system also can have multiple outputs (Fig. 5.3B), but each output must now include a filter capacitor sized in proportion to its load. Since a practical DC transformer would have a small filter capacitor on any output as shown in Fig. 4.17B, its value need only be increased to meet the requirement just mentioned. Total effective filter capacitance seen on the primary side of the transformer will appear to the buck SPC as if a single-output were being provided. Also, the load appearing at the output of the buck SPC is simply the parallel combination of the secondary loads reflected to the primary side of the DC transformer.

In the converters of Fig. 5.3, the DC transformer is current-fed and we realize

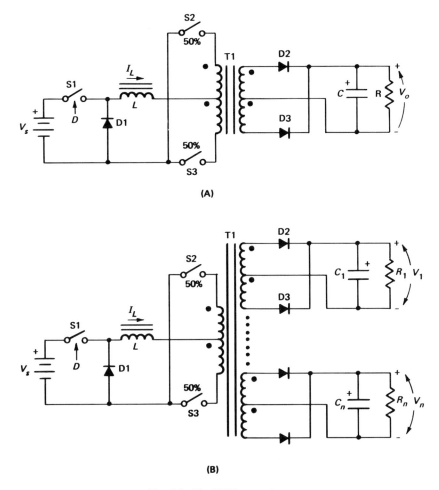

Fig. 5.3. The TRW converter.

the advantages of this type of source as outlined in Chapter 4. The well-controlled component stresses in these circuits make them particularly attractive for high power, multiple output applications.

There is no inherent requirement in the SPCs of Fig. 5.3 for the switching action of S1 to be synchronized with those of S2 and S3. They can be operated asynchronously, both in phase and frequency if desired. However, if S1 is operated at twice the switching frequency of S2 and S3 and locked in time phase with them, some advantage from a component stress standpoint can be gained, as shown by the circuit waveforms of Fig. 5.4.

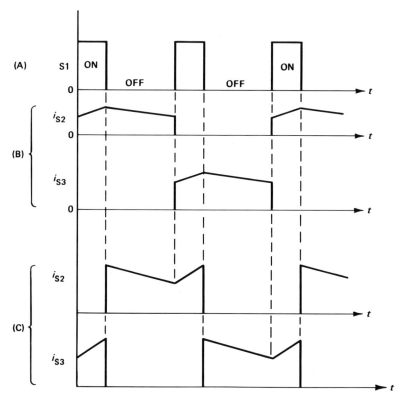

Fig. 5.4. Current waveforms for synchronized switching in the TRW converter.

Two synchronized possibilities are illustrated, switches S2 and S3 may be turned ON when S1 is turned ON or they may be turned ON when S1 is turned OFF. The resulting switch current waveforms are shown in Figs. 5.4B and C, respectively. The most desirable waveform is that shown in Fig. 5.4B, since this method of synchronization allows S2 and S3 to switch when the inductor current is at a minimum. If τ_L is not much greater than τ_{LC}, then the current stresses and corresponding switching losses in S2 and S3 can be quite small. This reduction in stress can be very helpful for applications where high output power is needed, or when high-frequency switching rates are required.

A small refinement could be added to the idealized waveforms of Fig. 5.4 so as to reflect potential overlaps in S2 and S3 conduction times. When both of these switches are ON, inductor current (I_L) divides between them. In this way, current levels are close to $I_L/2$ when the switches turn ON and OFF, which further reduces their switching losses.

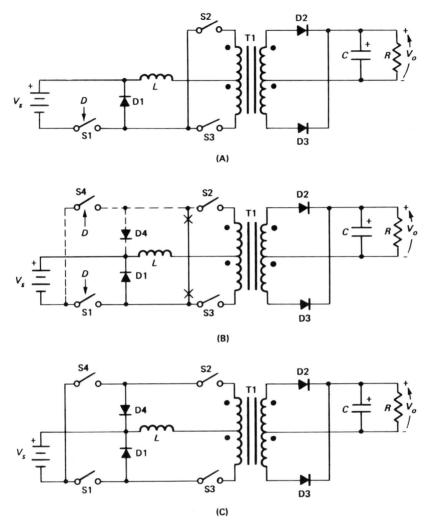

Fig. 5.5. Evolution of the IBM circuit.

The buck-derived converter of Fig. 5.3A can be modified in a number of interesting ways. One modification is shown in Fig. 5.5C, with intermediate evolutionary steps illustrated in Figs. 5.5A and B. In Fig. 5.5A, S1 is moved to the return point of the source voltage. Obviously, this move does not alter the switch or SPC function. The second step, shown in Fig. 5.5B, duplicates the original switch functions of S1 and D1 with the addition of S4 and D4. This step is completed by the removal of the connection between S2 and S3, which results in the converter system of Fig. 5.5C. Superficially, this "new" SPC would not seem to offer any advantages over that of the one shown in Fig. 5.3A, but a closer examination of it reveals that there are some operational differences, particularly in how the current in D1 and D4 commutates during a switching cycle of the system.

At the beginning of a switching cycle, S1 and S3 in Fig. 5.5C are both ON. Energy is stored in the inductor L and also delivered to the output load of T1 via diode D2. Part way through the switching period, S1 is turned OFF and the inductor current commutates to diode D1. Halfway through one complete switching cycle, the inductor current is then transferred from S3 and D1 to switches S2, S4, and diode D4. Because this is a current-fed converter system and the current through inductor L must be continuous (if operating in the continuous mode), switches S2 and S4 are closed before turning S3 OFF (remember—S1 is already OFF). The resulting reverse voltage appearing across D1 at this time will be smaller than that seen in the commutating diode of a normal buck SPC or its offspring shown in Fig. 5.3A. Lower reverse voltage will reduce the recovery time current transients through D1 as well as diode D4 on the next half cycle of converter operation. In a high-frequency or high-power converter application, these reduced current transient levels can significantly reduce electrical stresses on contemporary switch elements, lower commutating diode losses, and increase overall efficiency.

The SPC circuits in Figs. 5.2, 5.3, and 5.5 are very useful if their source voltage values are above 50 VDC. However, for lower source voltages, the conversion efficiency suffers because the inductor current must flow through two series switch elements during a substantial part of the switching cycle. This particular power loss problem can be eliminated by modifying the converter circuit of Fig. 5.5C, as shown in Fig. 5.6A. Here, circuit connection points for switches S1 and S4 are changed to the opposite ends of S2 and S3. The resultant switching sequence remains as described earlier, except that inductor current flows through only one switch element at any time in the switching cycle. The price paid for the elimination of one series element voltage drop is an increase in the peak OFF voltage across S1 and S2 from V_s to $(V_s + 2V_r)$, where V_r is the reflected load voltage through transformer T1. Since the SPC Fig. 5.6B is intended for low input voltage applications, this increased switch voltage stress is seldom a problem and represents a good trade.

The converter system of Fig. 5.6B is very useful, but it does have design drawbacks. First of all, it is relatively complex, using four switch elements and two primary commutating diodes. Second, the inductor current flows through both a switch and a diode during a portion of the switching cycle, which increases the power losses for low values of source voltage. As will be shown in Chapter 8, these losses can be reduced by using a tapped winding on transformer T1. How-

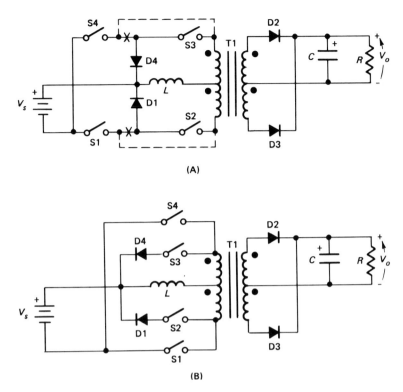

Fig. 5.6   Evolution of the Cronin circuit.

ever, a much better solution would be a reduction in the number of switches and commutation diodes.

Another buck-derived circuit exists which is very similar in structure to the one in Fig. 5.6B. This variation is illustrated in Fig. 5.7. Note that primary diode and switch connections are the same, but now a series inductor has been added on the secondary side of the converter. For multiple outputs from this converter, it is necessary to use a similar inductor on each individual output, as will be necessary for the quasi-squarewave converter to be discussed later in this chapter. This SPC circuit variation does not appear to have any advantages over the one shown in Fig. 5.6B and, in fact, has some significant disadvantages, the most serious one being the need for a multiplicity of inductors for multi-output applications. All such inductors must remain in a continuous current mode if reasonable cross-regulation of converter outputs is to be realized.

By making the inductor a two-winding element and repositioning the converter switches, the converter of Fig. 5.8 can be derived from the one shown in Fig. 5.6. This new circuit does not appear to have any advantages over its predecessor,

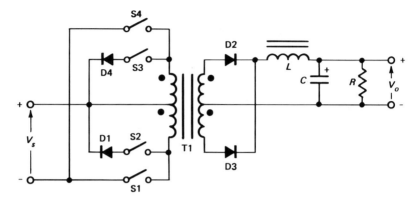

Fig. 5.7. The Hunter circuit.

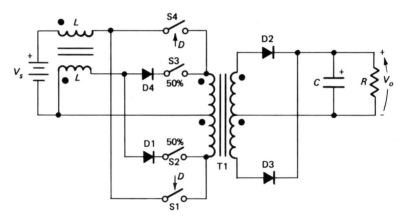

Fig. 5.8. The Severns circuit.

except that its structure is not patented and therefore lies in the public domain. It does, however, provide a conceptual stepping stone to another quite useful buck-derived converter, shown in Fig. 5.9A.

If S2, S3, and the secondary winding of inductor L in Fig. 5.8 are connected to the secondary side of T1 and the turns ratio of the windings on L made equal to the turns ratio of T1, then it is apparent that switches S2, S3 and diode D4 will no longer be needed and can be eliminated, as indicated in Fig. 5.9A.

We have now arrived at a current-fed converter circuit which uses only two primary switch elements with the inductor currents flowing through either a primary switch *or* a secondary diode D1, but *not* both simultaneously.

For low input voltage ranges, moderate winding transformation ratios and one or two outputs, the converter circuit of Fig. 5.9A is both simple and efficient.

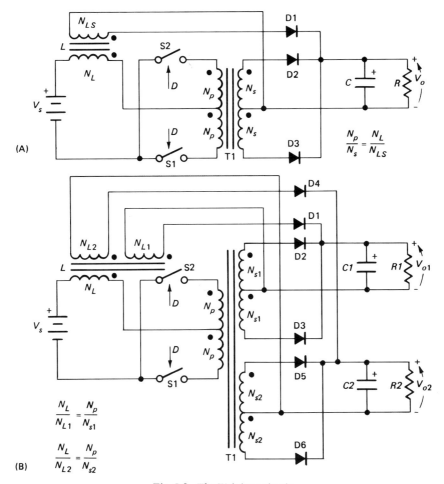

Fig. 5.9. The Weinberg circuit.

However, if large turns ratios are necessary between the windings of T1 and L, then switch voltage stress problems will arise from inherent primary-to-secondary leakage inductances. This can be a very serious problem at high output power levels for reasons mentioned in Chapter 4. To accommodate multiple outputs from this converter, inductor L must also have multiple windings and associated diodes as illustrated in Fig. 5.9B. When more than two outputs, high power delivery, or large transformation ratios are required, the converter circuit of Fig. 5.7 may be a better application choice.

There is another potentially serious problem which can occur in the converter circuit of Fig. 5.9A. Non-ideal transformers have finite magnetizing inductances.

Therefore, during the intervals when both primary switches are OFF, the voltage across both primary transformer windings cannot be zero because transformer magnetizing current is discharging through one of the output diodes, thus clamping the secondary winding voltage to that of the output load. This implies that during concurrent OFF intervals of the primary switches, an additional voltage above that of the source *plus* the reflected output voltage from L will appear across one of the switches, raising the OFF voltage level across one and reducing the OFF voltage level across the other. In addition, the residual flux in transformer T1 may increase because of the impressed secondary voltage, in turn increasing the chances of core saturation. Fortunately, there is a rather simple cure for this problem, which is shown in Fig. 5.10. Here, an additional secondary diode is added in series with the inductor. Both inductor commutating diodes are then connected so as to force the transformer secondary voltage to zero during the interval when both primary switches (S1 and S2) are in an OFF state.

When input-to-output DC isolation is not required and when the converter output voltage is higher than its input voltage, a direct-coupled version of the circuit of Fig. 5.9A may be used. Two examples of this type of converter circuit are given in Fig. 5.11. Note that the primary inductor is now tapped, and for different values of transformer turns and inductor connections, boost voltage gain can be obtained. Adding the transformer demagnetization feature shown in Fig. 5.10 to the converters of Fig. 5.11 is accomplished by a separate winding on the transformer, as illustrated in Fig. 5.11C.

We can now return to Fig. 5.1 and insert the DC transformer at positions C-C' and D-D'. The result is shown in Fig. 5.12A and B, respectively. From an inspection of these two SPC circuits, it can be seen that the commutating diode (D1) is no longer required in either case, since diodes D2 and D3 will pro-

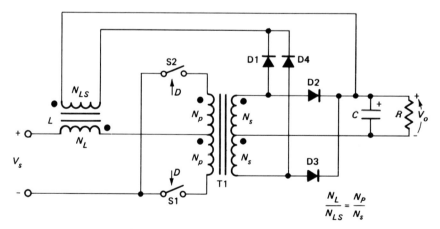

Fig. 5.10. An improved Weinberg circuit.

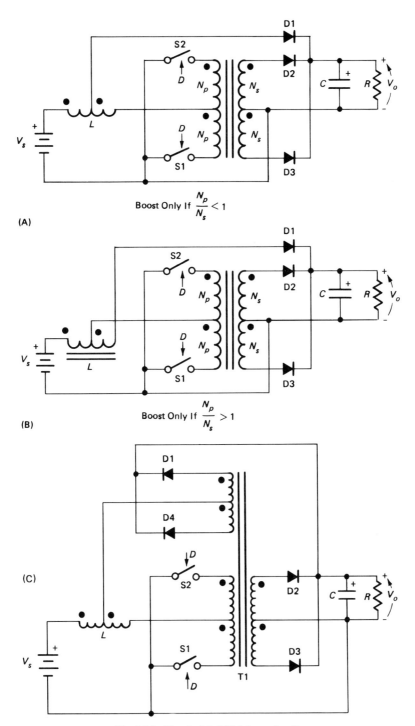

Fig. 5.11. Non-isolated Weinberg circuits.

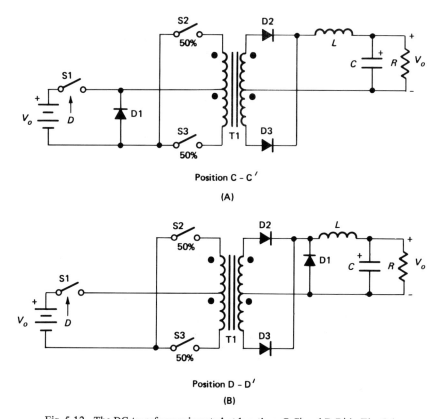

Fig. 5.12. The DC transformer inserted at locations C-C′ and D-D′ in Fig. 5.1.

vide this function of assuring continuous inductor current when switches S2 and S3 are both OFF. The two converters are therefore equivalent in operation. Occasionally, D1 is not removed from Fig. 5.12B in high output current applications. The reason for not removing D1 in this case lies in the fact that, during the converter intervals when S2 and S3 are OFF, the inductor current may commutate through D1 instead of through the two halves of the center-tapped transformer secondary winding via diodes D2 and D3. Therefore, the associated power losses in the secondary winding are reduced during inductor current commutation intervals.

Looking again at Fig. 5.12 we can see that the duty cycle control function of S1 could be performed by primary switches S2 and S3, since these latter switch elements are directly in series with S1 on alternating half-cylces. If duty cycle control of power conversion is moved to S2 and S3 (with S1 removed), the circuit in Fig. 5.13A emerges. The common name given to this circuit is the *parallel quasi-squarewave converter* and it is one of the most popular DC-DC conversion

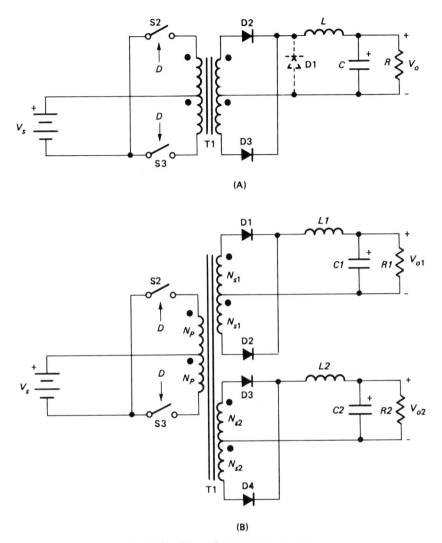

Fig. 5.13. The quasi-squarewave converter.

systems in use today. In applications requiring multiple DC outputs, additional transformer windings and filter networks can be added, as shown in Fig. 5.13B for a two-output case. Notice here that each DC output must have low-pass output filter. In this circuit, it is very possible for one output inductor to be operating in the continuous mode, with the other operating in the discontinuous mode! The character of any associated small-signal model of this SPC will depend on which converter output is being regulated. In normal operation, all filter inductors are usually maintained in the continuous mode to assure good cross-regulation between the regulated output and other unregulated ones. Any output of this SPC that drops out of the continuous mode will lose this highly desirable cross-regulation action. The problem of cross-regulation, as well as the number of output inductors required for multiple-output applications, are major draw-backs of this particular circuit approach. Another potential disadvantage is the fact that the DC transformer is now voltage-fed and therefore subject to transient currents produced by primary switch conduction overlap, transformer core saturation, and the other problems discussed in Chapter 4.

The final step in our evolutionary process is to insert the DC transformer at position E-E' in Fig. 5.1. The resulting SPC is shown in Fig. 5.14A for a single-output application, and in Fig. 5.4B for multiple-output situations.

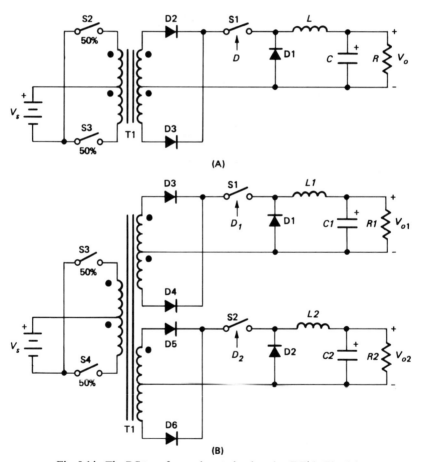

Fig. 5.14. The DC transformer inserted at location E-E' in Fig. 5.1.

This particular converter circuit is occasionally used and allows each of the outputs to be individually regulated by proper duty cycle control of the secondary switches (S1, S2 in Fig. 5.14B). The disadvantage of this circuit is the multiplicity of output switch elements. Because of their presence, the output switches add additional power losses in the converter, especially if the outputs are required to deliver high currents at low DC voltages.

A very common requirement encountered by designers today is the need for an SPC with one low-voltage (2-5 VDC) high-current output along with one or more higher-voltage (10-28 VDC) outputs with lower output current levels. If good voltage regulation and efficient conversion of power are required for all outputs, then the converter in Fig. 5.14B can be combined with the quasi-squarewave system of Fig. 5.13A. This possibility is shown in Fig. 5.15 for a two-output SPC. Here, the primary switch control loop is closed around the low-voltage high-current output, (V_{o2}). A separate switch (S1) and associated controls are used in the remaining secondary network to provide good regulation of the second DC voltage (V_{o1}). For this SPC to work properly, AC voltages

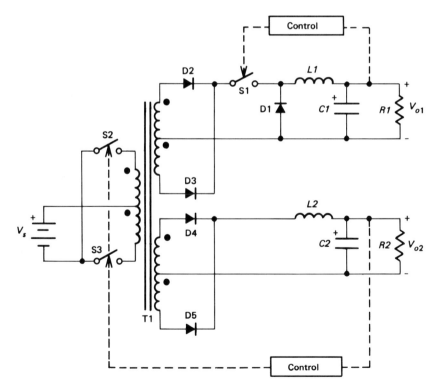

Fig. 5.15. The quasi-squarewave converter with an auxiliary post-regulator control system.

across the second output winding of the transformer must be made large enough so as to give the second control switch (S1) reasonable duty cycle range for regulation of voltage V_{o1}. Since the SPC's primary switches are also duty cycle controlled, the AC voltages across the secondary of the second winding are regulated by control of V_{o2}; therefore, the duty cycle of S1 will remain close to 90% or better as the SPC's source voltage (V_s) changes as long as the inductors remain in the continuous mode.

5.2 FULL- AND HALF-BRIDGE DC TRANSFORMERS IN THE BUCK SPC

A problem common to all of the converters in Figs. 5.12 through 5.15 is that all primary switches (S2 and S3) will see a peak OFF voltage stress of twice the source voltage value (or greater, when transients are considered). These voltage stress levels can be reduced by using either a full or half-bridge DC transformers, rather than the parallel DC converter were used up to this point.

In the previous section, if a full-bridge DC transformer had been chosen in our evolutionary investigations, we would find that identical manipulations could be applied and similar circuits derived. Figs. 5.16 and 5.17 show the full-bridge

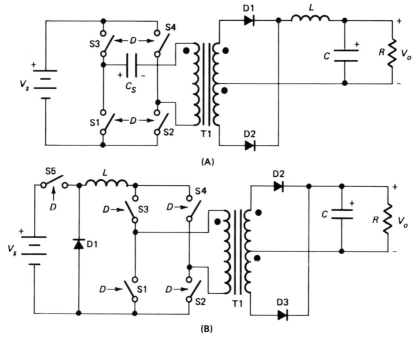

Fig. 5.16. Full-bridge buck-derived variations (I).

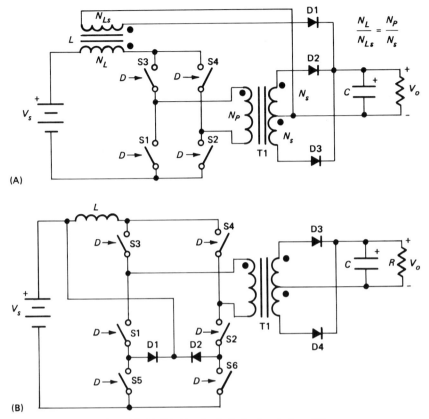

Fig. 5.17. Full-bridge buck-derived variations (II).

equivalent for the circuits of Figs. 5.16, 5.12, 5.9A, and 5.5C, respectively. Full-bridge versions of circuits are used when the need arises for high output power levels (>500 W) at high values of source voltage. From our full-bridge discussions in Chapter 4, the switches see a reduced peak voltage stress when they are in an OFF state. Here again, the presence of additional voltage stresses from inherent parasitic circuit elements has been ignored.

For lower power levels and high source voltages, the half-bridge versions, such as those shown in Fig. 5.18, are preferred.

Both the full and half-bridge SPCs call for an untapped single primary winding on their transformers. From a copper utilization standpoint, a single primary winding is advantageous and helps to reduce transformer manufacturing cost and increase its efficiency. In addition, the need to obtain good magnetic coupling between the two halves of the primary winding of a parallel transformer is eliminated.

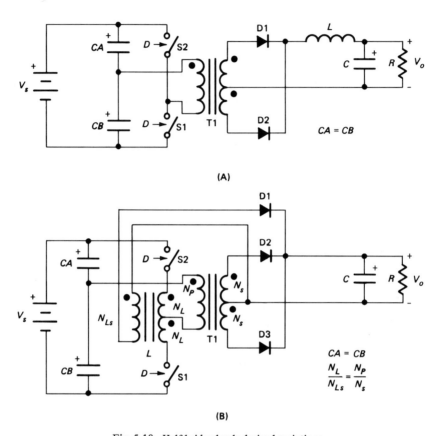

(A)

(B)

Fig. 5.18. Half-bridge buck-derived variations.

In Figs. 5.16 and 5.17, all of the primary switch elements are shown to be modulated in duty cycle. This is a usual mode of operation but not a mandatory one. Switches S1 and S2 or their counterparts S3 and S4 could be operated with fixed conduction duty cycles of 50%, with variable duty cycle modulation applied to the remaining switch pair. In some SPC applications, a simplification of switch drive circuitry will result. Alternatively, switches S1 and S3 or S2 and S4 may be operated at fixed 50% conduction duty cycles and duty cycle control applied to the other two switch elements.

5.3 THE SINGLE-ENDED DC TRANSFORMER IN THE BUCK SPC

Simplicity of topology is a very great virtue in a converter since it usually reduces cost and weight while increasing reliability. For this reason, converter circuits which use only one switch element are very popular among designers.

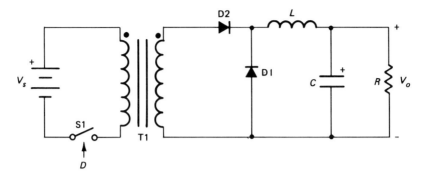

Fig. 5.19. The forward converter.

The single-ended form of the quasi-squarewave converter is shown in Fig. 5.19. This circuit is commonly known as the *forward converter* and is widely used in Europe. Applications of the forward converter using a single BJT switch and having an output power of 2 kW or more do exist, although this SPC is normally used at much lower output power levels. It is debatable for many practical reasons if this SPC is really advantageous at very high power levels.

As was pointed out in Chapter 4, a practical single-ended DC transformer circuit must have some provision for transformer core reset. If we apply the reset schemes developed in Section 4.2 to the forward converter of Fig. 5.19, a wide variety of related circuits can be evolved. A few of the resulting converter possibilities are shown in Figs. 5.20 through 5.24.

If a physically smaller transformer with better core utilization properties is desired, then current reset schemes can be used. Three possible methods are illustrated in Fig. 5.22. Reset using a tertiary winding and a current source (Fig. 5.22A) is fairly common; however, this approach can be readily accomplished by using the current source of the output filter, namely, inductor L!

As shown in Fig. 5.22B, the current in inductor L may be used as a means of core reset by simply providing a tap on the transformer secondary and a diode (D1) for connection to the inductor. The transformer core may also be reset by using a field bias from a built-in permanent magnet, as shown in Fig. 5.22C. Biased cores are commercially available and usually made of ferrite materials with inherent permanent magnets.

There is one mechanism missing from the SPC circuits in Fig. 5.22, namely, a means for controlling the amplitude of the primary voltage during core reset intervals. Some limits must be placed on these amplitudes to protect the SPC switch element from possible voltage breakdown. Any of the voltage clamp schemes of Chapter 4 may be used, one example of which is shown in Fig. 5.23.

It might appear that the choice of reset schemes is an arbitrary one, but such is not the case. The reset scheme chosen will determine the maximum duty cycle

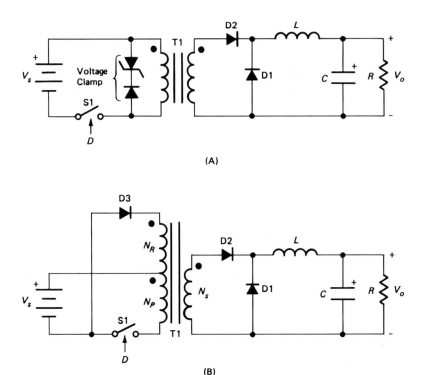

Fig. 5.20. Forward converter core reset variations (I).

limit of primary switch control which, in turn, determines the peak switch OFF voltage level and ON current value.

Since the forward converter of Fig. 5.19 is buck-derived and therefore has the properties of a buck SPC, the maximum value of switch duty cycle will occur at the lowest source voltage and highest load value. As V_s increases, the duty cycle (D) of the primary switch (S1) will decrease. The larger D can be made, the smaller the peak switch current will be, and the power level which can be controlled by a given switch will be greater.

Other reasons for maximizing switch duty cycles are the resulting reductions in the physical size of inductor L and the rms current in any input filter capacitor across the source voltage. In an elementary buck SPC, as its source voltage approaches minimum value, the switch duty cycle could approach unity in order to maintain output voltage regulation. However, in the forward converter, some time is always required to reset the core of its transformer. We can make D larger by making V_s larger, but we must be careful not to exceed the breakdown rating of the switch when V_s goes to its highest value. An ideal voltage

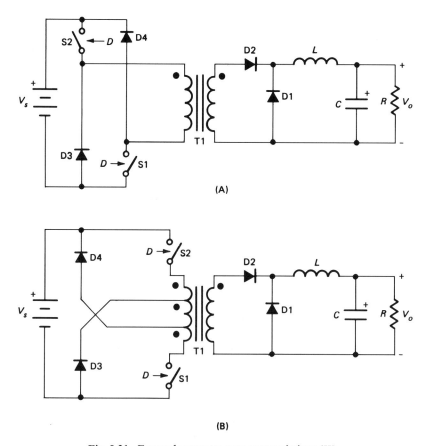

Fig. 5.21. Forward converter core reset variations (II).

reset scheme would provide a high reset voltage when V_s is low and a lower reset voltage when V_s is high. This would allow the maximum possible value for D at the lowest value for V_s for a given voltage rating of the switch. Unfortunately, this ideal reset scheme is not easily realized and most circuits are a compromise. The best compromise is to use a constant reset voltage such as that shown in Figs. 4.5 and 4.6B. The variable reset scheme given in Fig. 4.6A is the worst, since the *lowest* reset voltage occurs at the *lowest* value of V_s—just the opposite of what is desired.

In practice, limit values of D from 60 to 70% are quite common. Compared to a parallel DC transformer quasi-squarewave converter (Fig. 5.13A), the forward converter topology uses only one switch element and requires fewer windings on its transformer. On the other hand, the inductor and transformer of the

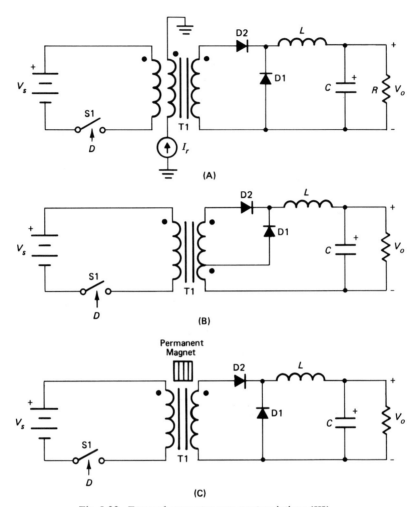

Fig. 5.22. Forward converter core reset variations (III).

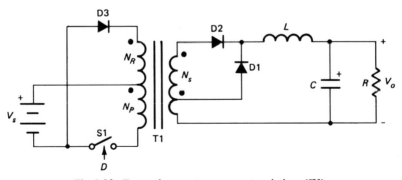

Fig. 5.23. Forward converter core reset variations (IV).

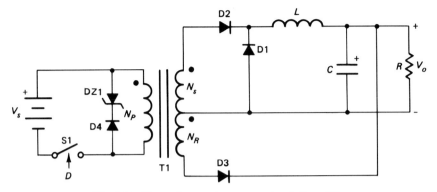

Fig. 5.24. Forward converter core reset variations (V).

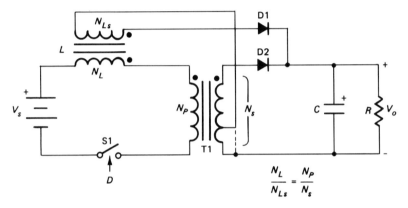

Fig. 5.25. An example of a single-ended Weinberg converter.

forward SPC are usually physically larger than their contemporaries in a symmetrical quasi-squarewave SPC.

Single-ended transformer equivalents are not limited to reduced forms of the quasi-squarewave SPC, although the forward converter just discussed is the most popular and in wide use today. A single-ended version of the SPC circuit of Fig. 5.9A is shown in Fig. 5.25, as an example of another circuit form which uses a different parent. By applying the various core reset schemes described for the forward converter, another generation of single-ended SPC's based on the converter of Fig. 5.25 would result.

6. Boost-Derived Circuits

In this chapter, we will essentially repeat the evolutionary processes of Chapter 5, using the basic boost SPC of Chapter 3 as our starting point. As a result, a large family of boost-derived SPC circuits will be created.

It is interesting to note that most of the buck-derived circuits of Chapter 5 were developed independently by many individual designers over the past twenty year period. However, the processes used to demonstrate electrical relationships between them were developed after many buck-derived converters were in more or less popular use. Such was not the case for boost-derived SPCs with relatively few examples of derivative converters to be found only in recent literature. In this chapter, the procedures which we used to show the relationships among buck-derived SPC circuits become synthesis tools to develop new boost-related SPCs. In many cases, the resulting SPCs will be unusual and interesting in form. Many of the boost-derived circuits to be discussed arise from repeating simple manipulation techniques that were applied to the buck-derived circuits of Chapter 5.

All of the circuits to be developed in this chapter retain the basic properties of the elementary boost converter with the exception that DC isolation, arbitrary input-to-output voltage ratio, and multiple-output capability are now possible.

6.1 PARALLEL DC TRANSFORMERS WITHIN BOOST CONVERTERS

Five logical points to insert a DC transformer element into the basic boost SPC are shown in Fig. 6.1.

Insertion of the DC-DC transformer at A-A' results in the converter of Fig. 6.2, a pre-regulated DC-DC converter with a boost SPC as the pre-regulation circuit. This converter can be advantageous if its source voltage has a low value (e.g., 12 VDC mobile applications) because the DC transformer switches can operate at higher voltage and lower current levels. In general, for low values of input voltage, this circuit will be more efficient than its buck-derived counterpart of Chapter 5.

When the DC-DC transformer is inserted at location B-B' or C-C' of Fig. 6.1,

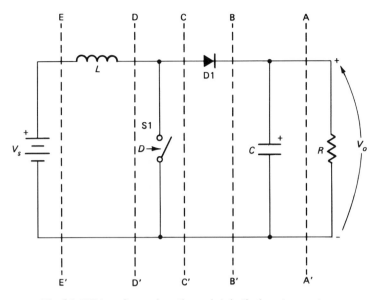

Fig. 6.1 DC transformer insertion points in the boost converter.

the equivalent converters of Fig. 6.3 appear. Again, we have a case where D1 is not required for the circuits to function properly, and therefore can be omitted, as illustrated in Fig. 6.4A.

The SPC circuit in Fig. 6.4A is a viable conversion approach, but it does use three switch elements. If the duty cycles of S2 and S3 are made variable and always greater than 50% (i.e., if S2 and S3 have *overlapping* conduction intervals)

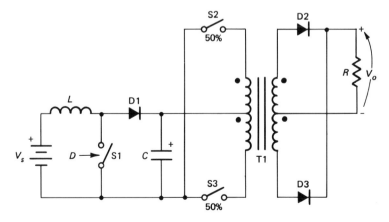

Fig. 6.2 The DC transformer inserted at location A-A' in Fig. 6.1.

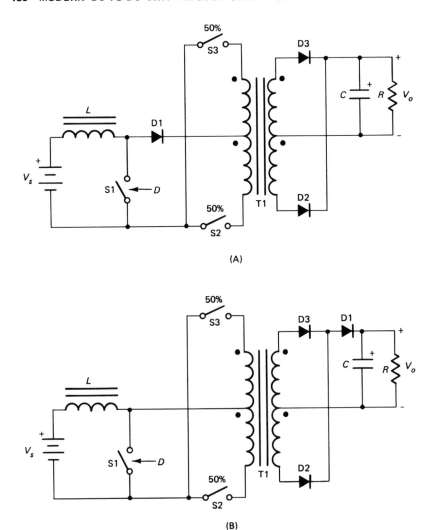

Fig. 6.3 The DC transformer inserted at locations B-B' and C-C' in Fig. 6.1.

so as to permit the primary of the transformer T1 to present a short circuit to inductor L for a portion of each half cycle), then S1 is no longer needed and it can be eliminated as shown in Fig. 6.4B. Fig. 6.4B also gives an indication of how easily a need for multiple outputs can be accommodated. One of the advantages of the overlapping primary switch conduction is the equal division of inductor current between S2 and S3, thus reducing their average and rms current

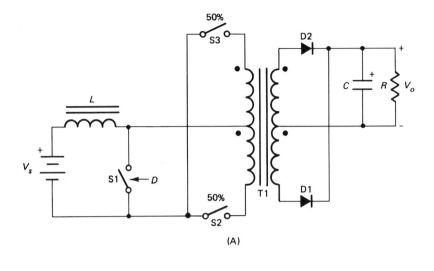

(A)

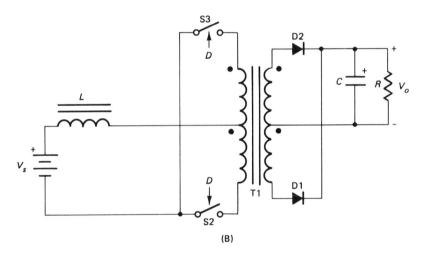

(B)

Fig. 6.4. The Clarke circuit.

levels as well as the primary winding rms current magnitude. When MOSFET devices are used as the switch elements, a significant improvement in conversion efficiency can be achieved here.

Proceeding to position D-D′ of Fig. 6.1 and inserting the DC transformer produces the converter circuit of Fig. 6.5A, in which the shunt switch now appears on the secondary side of the transformer. Having this duty-cycle modulated switch in this position can be very advantageous, if strict DC isolation is required between primary and secondary circuits as well as for associated voltage regulation controls. Here, any regulation control circuitry can share a common reference with the output voltage being monitored.

The converter in Fig. 6.5A does have one potential problem. If the SPC's output is required to deliver a low DC voltage at high current levels, then having D1 in series with the rectifiers (D3 and D2) can cause unacceptable power losses. In addition, the ON current of S1 will be large, further increasing such losses. This problem can be fixed quite easily, as shown in Fig. 6.5B, where the clamp diodes (D1, D2) along with the secondary switch are operated at higher voltage and lower current levels, with the converter's output current flowing only through each of the rectifier diodes (D3, D4).

This idea of secondary-referenced duty cycle control can be extended, as shown in Fig. 6.5C, by placing S1 on a separate auxiliary winding. This provides not only DC isolation between the primary and both secondaries, but also between any voltage regulation control switch element! The auxiliary winding also allows the voltage and current levels in S1 to be optimized to suit the capabilities of the device used to realize the switch function. As shown earlier for the converter of Fig. 6.4B, auxiliary transformer windings can also be added to produce multiple outputs if desired.

In the converters of Fig. 6.5, it is not mandatory that S1, S2, and S3 operate synchronously, but it is definitely advantageous. If S1 operates at twice the switching frequency of S2 and S3, and if S1 and S2 or S1 and S3 are turned ON simultaneously, then S2 and S3 will turn ON and OFF at inductor current minimum points, much as was observed for the current-fed buck converters in Fig. 5.3.

If the core of the transformer is made from a material with a high-B_r characteristic (often termed a "square-loop" material), then a number of additional boost-derived circuits become possible. In the converter shown in Fig. 6.6A, the conduction of duty cycle of S1 and S2 is fixed, usually just over 50% for slight overlaps. The switching frequency (f_s) is the control variable for regulation and is adjusted so that transformer T1 goes into saturation on each half cycle of the switching sequence. Saturation of T1 emulates to a great degree the same "shorting" function as the shunt control switch of Fig. 6.4. As f_s is made lower, the time spent by the transformer primary in a "shorted" condition over a half-cycle interval ($T_s/2$) will become greater. Therefore, the output voltage (V_o) is inversely proportional to the value of f_s. The converter of Fig. 6.6A works very well as long as switching frequency is kept low enough so that the transformer core can be driven from $+\phi_s$ to $-\phi_s$. As the switching frequency is raised, a point will be reached where core temperature rise from power losses may become excessive, at which time $\Delta\phi$ must be reduced. Keep in mind that the trans-

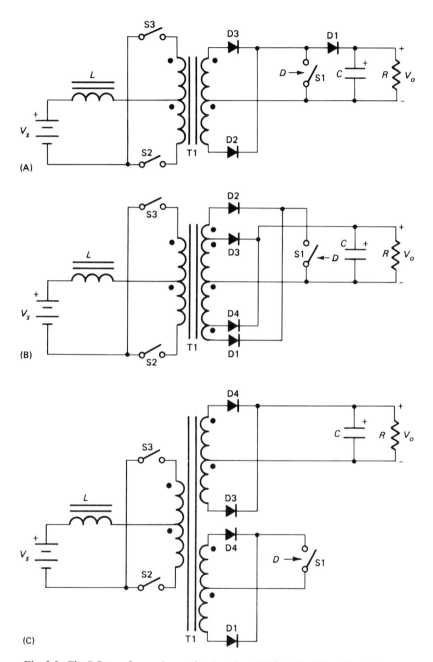

Fig. 6.5. The DC transformer inserted at locations D-D' in Fig. 6.1 with variations.

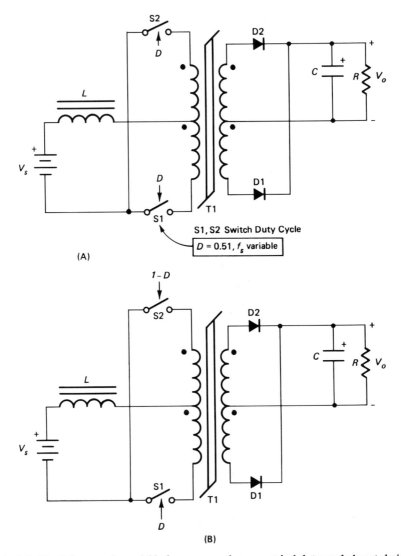

Fig. 6.6. Fixed duty cycle, variable-frequency and asymmetrical duty cycle boost-derived converters.

former core material chosen will place an upper boundary on the switching frequency.

Alternative converter circuits which exploit the saturable properties of a square-loop core material allow for arbitrary changes in flux and operate at a fixed switching frequency, as shown in Fig. 6.6B and Fig. 6.7. In the SPC of Fig. 6.6B,

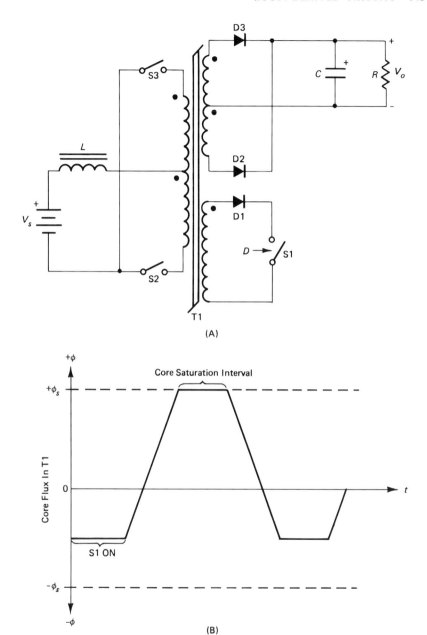

Fig. 6.7. A secondary shunt-switch saturable-core converter circuit.

the duty cycles of S1 and S2 are deliberately made asymmetrical so that during each full switching cycle the transformer core saturates, as it cannot support any significant asymmetry in primary volt-second levels. Here it is not necessary that both primary windings be identical in number of turns, but reasonably good magnetic coupling between them is still highly desirable. By making the windings asymmetrical, the two switch duty cycles of conduction can be altered to more favorably distribute the conversion power losses between them.

The circuit in Fig. 6.7A is rather unusual in that, on alternating half cycles, the primary "shorting" action is provided first by the closure of switch S1 and then by saturation of the core of T1. A diagram showing typical core flux excursion as a function of switching times is illustrated in Fig. 6.7B.

One of the simpler boost-derived SPCs from an overall circuit point of view is shown in Fig. 6.8. Here, switches S1 and S2 are operated at a fixed duty cycle and the saturation time of transformer T1 is controlled by the magnitude of the bias current source, I_c. This converter eliminates the need to modulate the primary switch conduction intervals. The modulation action is performed by the transformer alone, lending considerable simplification to the SPC's control circuitry. To adjust the gain due to I_c, N_c can be varied. It may also be advantageous to make the fixed duty cycles of S1 and S2 asymmetrical. Alternatively, transformer turns N_1 and N_2 may be made asymmetrical. Both of these alter-

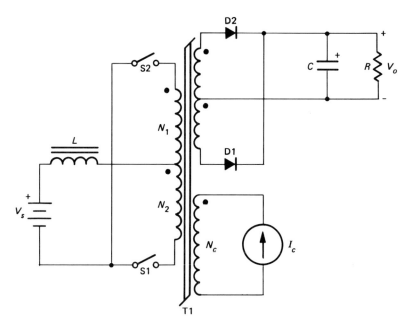

Fig. 6.8. A current-controlled boost-derived converter.

natives would have the effect of supplying a fixed DC bias for T1 and allowing the operating point of I_c to be adjusted accordingly in an optimum fashion.

Many boost-derived current-fed converters have an inherent problem when their switching action is shut down. For example, if drive power is suddenly removed from all converter switches, what happens to the stored energy in the inductor L? In many cases, in the absence of protective circuit arrangements, this energy will be dissipated by avalanche breakdown of the converter's switch elements. For BJTs, in particular, this can be a fatal situation. Two commonly used methods to circumvent this energy removal problem are shown in Fig. 6.9.

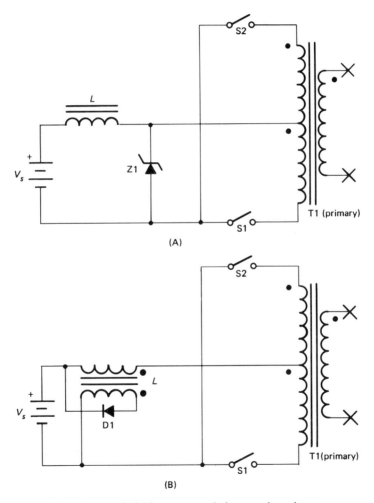

Fig. 6.9. Boost-derived converter switch protection schemes.

In Fig. 6.9A, residual energy of the inductor at converter shutdown is dissipated in a shunt power zener diode. Since this zener needs to absorb only the energy in L on a one-time basis (at converter turnoff), it does not have to dissipate much average power and can be quite small. An alternative measure is shown in Fig. 6.9B, where a secondary winding has been added to inductor L. Thus any residual energy will be returned to the SPC's voltage source. The turns ratio of this auxiliary winding relative to the primary is adjusted so that the commutating diode (D1) does not conduct during normal converter operation. This condition usually implies that voltage magnitudes across converter switches could be higher than that produced by the clamp of Fig. 6.9A, especially if the source voltage is large at the instant of SPC system shutdown.

6.2 BRIDGE AND HALF-BRIDGE DC TRANSFORMERS WITHIN BOOST CONVERTERS

We can insert full and half-bridge DC-DC transformers within basic boost SPCs just as easily as was possible for the buck-derived converters of Chapter 5. Two circuit examples are shown in Fig. 6.10A and 6.10B.

Examination of the SPC circuit in Fig. 6.10A reveals that a variety of switch sequences can be accommodated without compromising conversion performance. Three of these sequencing possibilities are shown in Fig. 6.11. In Fig. 6.11A, all four switch elements have conduction duty cycles greater than 50%. Here, twice each switching cycle, all four switches are ON simultaneously. If MOSFETS are used as switch elements, then the inductor current will divide relatively equally between switch sets S1 and S3 and S2 and S4. This current-sharing feature reduces primary switch rms current levels, especially at low input voltages when switch overlap interval and inductor current are at their maximum values. When BJT switches are used, current-sharing may not be equalized to the same degree. Therefore, one pair of BJT switches may take more load current than the other pair.

In a full-bridge converter, at least two of the primary switches will require DC-isolated drives. The simplest and most common means for providing an isolated drive network is to use transformer-coupling between the drive source and the associated power switch element. In a buck-derived bridge converter, a single drive transformer with multiple-output windings (e.g., see the circuit of Fig. 4.15) can be used since switch duty cycle (D) of conduction is always less than 50%. However, in boost-derived bridge converters, D is greater than 50% and a separate drive transformer will be necessary for each DC isolated switch element. Additional drive networks complicate overall design and their number should always be minimized.

The switch drives can be simplified by using the timing sequence shown in Fig. 6.11B, where S3 and S4 are operated at a fixed D of 50% and duty cycle

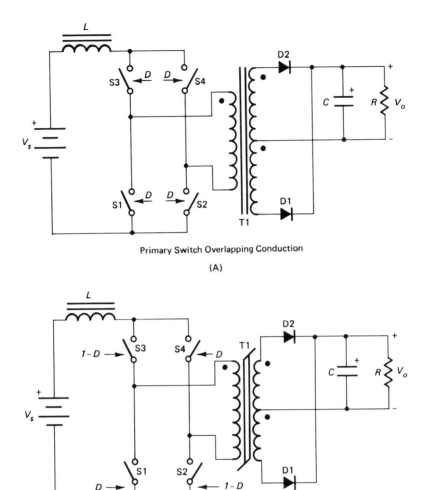

Primary Switch Overlapping Conduction

(A)

Asymmetrical Switch Modulation

(B)

Fig. 6.10. Full-bridge boost-derived converter variations.

control applied only to S1 and S2. This sequence simplifies the drive problem, but now only two SPC switches are ON during conduction overlap periods.

A further drive simplification is shown in Fig. 6.11C. Here all four SPC switches are operated with a conduction duty cycle of 50%, with the conduction phase of S1 and S2 varied relative to that of S3 and S4. This sequence produces a conduction overlap twice each complete switching cycle, once by S1 and S3

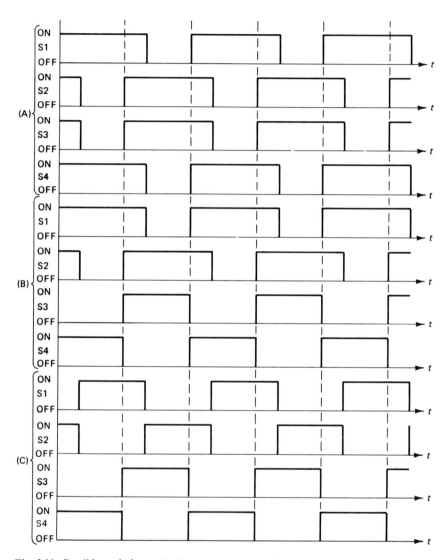

Fig. 6.11. Possible switch conduction sequences in the overlapping conduction boost converter.

and once by S2 and S4. Complete DC isolation of drive power for the four SPC switches can be achieved by using two transformers, each with two output windings. These drive transformers handle only low-power symmetrical signals, and are usually very small in physical size.

As shown by the examples in Fig. 6.12, half-bridge boost converters can be

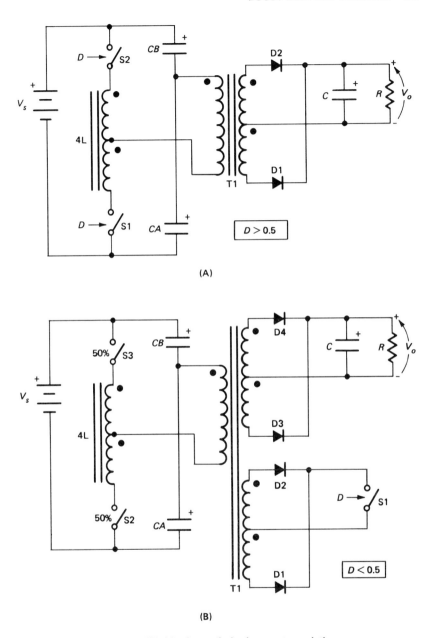

Fig. 6.12. Half-bridge boost-derived converter variations.

readily derived, but it is necessary that the input inductor L be center-tapped and inserted between the SPC switches.

6.3 SINGLE-ENDED DC TRANSFORMERS
WITHIN BOOST CONVERTERS

A single-ended DC-DC transformer can be combined with a basic boost SPC. Such combinations, however, result in converters that require *two* primary switch elements, rather than only one as was the case for buck-derived single-ended transformer designs.

The boost-equivalent of the buck-derived forward converter is shown in Fig. 6.13A. At the beginning of a switching cycle, S1 is ON and S2 is OFF. Inductor

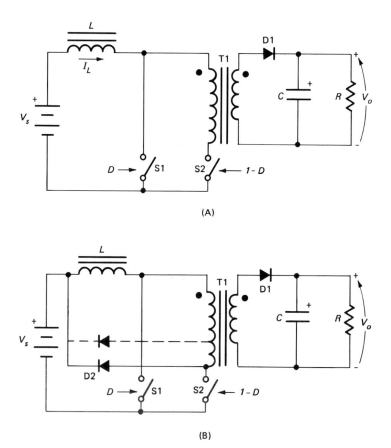

(A)

(B)

Fig. 6.13. A single-ended boost-derived converter.

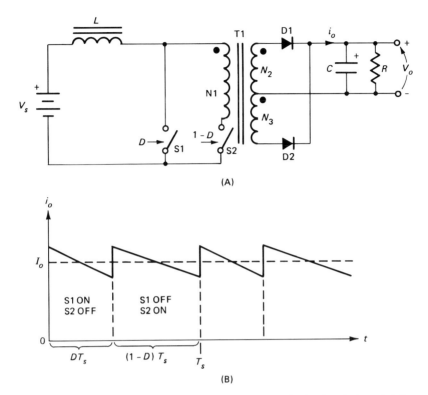

(A)

(B)

Fig. 6.14. A method for reducing the output current ripple in the single-ended boost converter.

current therefore flows through S1. With S2 open, energy to sustain the output voltage load comes from the filter capacitor C. When S1 turns OFF and switch S2 closes, it now allows the inductor current to flow through the primary of T1 and to deliver its stored energy to the output of the converter. As long as inductor current is flowing in a continuous fashion, one switch must always be closed for proper converter operation. A simple transformer core reset scheme for the converter of Fig. 6.13A is shown in Fig. 6.13B. Of course, other reset schemes discussed in Chapter 4 could also be used. The current of the inductor could also be used to reset the core of T1 by adding a small tertiary winding in series with switch S1.

Another interesting SPC circuit can be derived from the single-ended boost converter of Fig. 6.13A. Normally, the magnetizing inductance of T1 is made as large as possible to minimize the amount of energy stored in its core. However, if this magnetizing inductance is deliberately made relatively small and the stored energy of the core discharged into the secondary (as shown in Fig. 6.14A), then it is possible to greatly reduce the converter's output ripple current by judicious choice of the values of the inductor and T1's magnetizing inductance. The resultant output current waveform is illustrated in Fig. 6.14B. As shown, discontinuities in the waveform still exist at switching points but can be made smaller by optimizing the value of the magnetization inductance of T1. Another variable for controlling output ripple current magnitude is the turns ratios between N2 and N3.

7. Combinations of Converters

Up to this point, we have been discussing the two basic DC-DC converters and combinations of each with a single DC transformer. Despite these restrictions, many unusual and different SPC circuits were derived as well as some converter topologies that are in popular use today.

However, there are many useful SPC circuits that are combinations of more than one elementary converter. Some of the more complex SPCs also include a DC transformer or even a simple AC transformer to further enhance their utility. As we will see, the number of SPC circuit variations is limited only by the imagination and persistence of the designer, although the point of vanishing returns from increased circuit complexity inevitably puts a practical boundary on innovation.

The SPC building blocks available at this point in the discussion are the basic buck and boost converters, AC and DC transformers, plus the many buck and boost-derived circuits developed in Chapters 5 and 6. For simplicity in the discussions to follow, we will attempt to treat each of the combinational as interconnections of separate two-port electrical networks. A few of the interconnection possibilities of two-port networks are shown in Fig. 7.1. Obviously, a multi-output SPC or DC transformer system could be viewed as a multiport network. However, we will find plenty of complexity in the circuits of this chapter without including such viewpoints.

7.1 CASCADING IDENTICAL CONVERTERS

Perhaps the simplest combination of SPC's would be the cascade connection of identical converter topologies, such as the two buck converters shown in Fig. 7.2A. For this SPC operating in the continuous mode, its ideal overall voltage gain (M) will be:

$$M = D_a \cdot D_b. \tag{7.1}$$

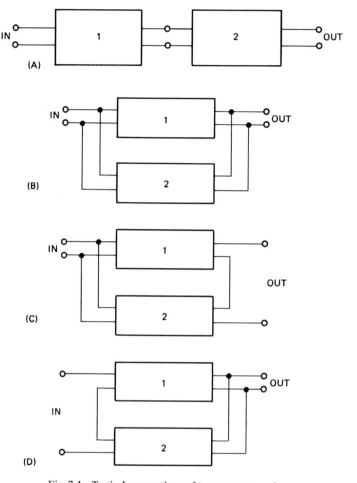

Fig. 7.1. Typical connections of two-port networks.

In some applications, D_a (or D_b) is held at a fixed value, while the other switch is varied to maintain output voltage regulation. In many cases, for simplicity of drive circuitry, duty cycle control of S1 and S2 are generated from a single source. In such cases, D_a will be equal to D_b so that

$$D = \sqrt{M}. \tag{7.2}$$

Remember that for a single-stage buck SPC:

$$D = M. \tag{7.3}$$

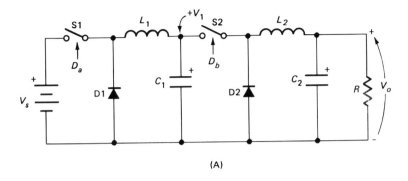

(A)

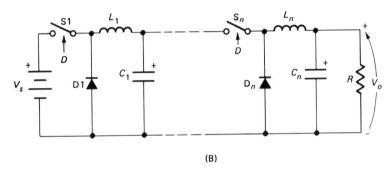

(B)

Fig. 7.2. Cascaded buck converters.

Since

$$M \leqslant 1 \tag{7.4}$$

the switch duty cycle for two buck stages in tandem will *always* be greater than that of a single stage for a given value of M. Several benefits become evident— first, since the average source current (I_s) through S1 must remain the same (same V_s and input power levels), then the peak current (I_b) through S1 must go down as D is increased. For example, if $M = 10$, then $I_b = 10 I_s$ for a single-stage buck converter operating in a continuous mode. However, for a comparable two-stage buck system, $I_b = \sqrt{10} I_s$. This implies that peak current levels of S1 will be much lower in the two-stage case and, therefore, it can control more power than an equivalent single-stage counterpart. Switch S1 and diode D1 will still see OFF voltage stresses equal to V_s, but S2 and D2 only have to sustain an OFF voltage (V_1) given by:

$$V_1 = \sqrt{V_o \cdot V_s} \tag{7.5}$$

for $D_a = D_b$. Since V_1 is always less than V_s in value, the OFF voltage capabilities of S2 and D2 are reduced accordingly.

If M is small (in the range of 0.4 or less), there is no real advantage in cascading the two converters. It would be better to simply implement the function of S1 with paralleled switch components for higher current-handling capability and remove the second low-pass filter network (L_2, C_2). On the other hand, if M is large, then the cascade connection of two buck SPCs should be considered for component stress reduction reasons.

This cascade principle can be extended to any number (n) of buck SPCs as shown in Fig. 7.2B. The ideal voltage gain of a complete system is then:

$$M = D^n \tag{7.6}$$

assuming all switches are operated at an identical duty cycle, D. From a practical standpoint, it is questionable that more than two buck stages in cascade offer any advantages. Nevertheless, cascade arrangement of many stages is still a viable concept.

For a number (n) of cascaded boost SPCs operating in the continuous mode, at same duty cycles of conduction, ideal voltage gain is given by

$$M = \frac{1}{(1 - D)^n}. \tag{7.7}$$

7.2 CASCADING DISSIMILAR CONVERTERS

Several useful and popular SPC circuits can be obtained from cascade combinations of buck and boost converters. For example, one can precede a boost converter with a buck converter as shown in Fig. 7.3A. This circuit can be greatly simplified by a bit of topology manipulation as illustrated in Figs. 7.3 and 7.4.

First, in Fig. 7.3B, we presume that S1 and S2 are synchronized and have identical duty cycles. Therefore, the functions of S1, S2, D1, and D2 can be replaced with an equivalent DPDT switch as indicated. Notice that capacitor C1 has been deliberately omitted in Fig. 7.3B. Up to this point in our discussion, the buck converter has always been shown with an output filter capacitor but, strictly speaking, it is not required. In theory, the inductor of a buck SPC can be made arbitrarily large so as to reduce load ripple current amplitudes without the need for additional capacitive filtering. In actual practice, however, it is much easier to add an output filter capacitor to keep the physical size of the inductor small. The output capacitor of the boost converter cannot be eliminated because its output current is always pulsating, regardless of how large its inductor is made. In the example of Fig. 7.3B, considerable ripple in L1 can be tolerated since L2 and C2 provide a second stage of low-pass filtering.

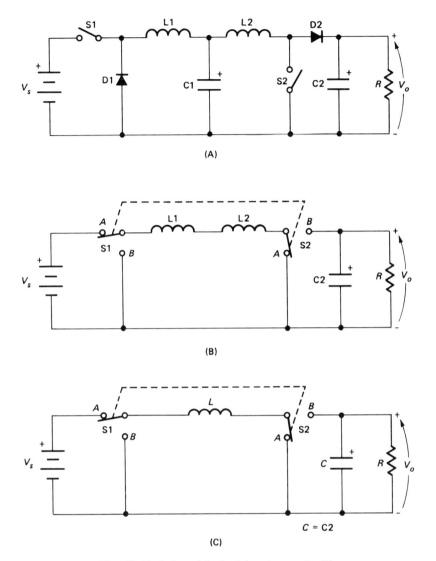

Fig. 7.3. Evolution of the buck-boost converter (I).

Having eliminated C1, inductors L1 and L2 can now be combined into a single element, as shown in Fig. 7.3C, since they are series-connected in one branch of the topology.

If we are willing to accept a polarity inversion of the output voltage, the circuit can be further simplified by rearrangement of the DPST switch/inductor network of Fig. 7.3C to that shown in Fig. 7.4A. A practical realization for this

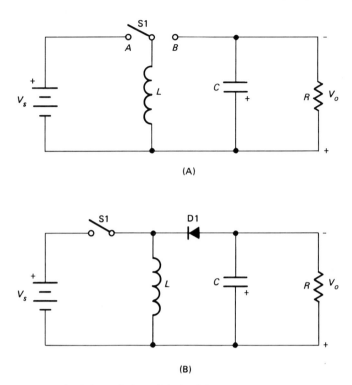

(A)

(B)

Fig. 7.4. Evolution of the buck-boost converter (II).

latter circuit is given in Fig. 7.4B and is commonly called a *buck-boost converter*. The buck-boost converter uses the same number of circuit components as does a buck or boost SPC and has an ideal voltage gain equal to $D/(1 - D)$. Therefore it is possible to have output voltage magnitudes that are either greater or less than that of the converter's input voltage.

However, there is a price to be paid for this versatility in voltage gain. The buck-boost SPC has the input properties of the buck converter and the output properties of the boost converter. Both the input and output currents of this converter are pulsating, which increases the sizes of additional input and output filters used for EMI reduction. In addition, its switch and diode currents are greater (for a given power level and voltage gain) than those of the basic buck or boost converters. For these reasons, the buck-boost SPC is rarely used in high DC power processing applications.

The buck-boost converter is considered by many SPC design engineers to be an elementary SPC form. As we have seen, this SPC uses the same number of circuit components as do buck or boost converters, and certainly is a viable DC conversion scheme. Also, it is possible to treat the network formed by S1, D1, and L in Fig. 7.4A as an essential SPC feature and, by processes of rotation and mirror-imaging, used to logically evolve buck or boost SPC topologies.

As we have seen, it is possible by a simple manipulation to derive the buck-boost converter topology from the cascade connection of the two elementary converters. The derivation of the buck-boost SPC small-signal model given in Chapter 10 also supports this view. For these reasons, we have chosen here to classify the buck-boost converter as a special combination of buck and boost SPCs, rather than an elementary converter unto itself. Since this assumption has proven to be self-consistent, we have retained it throughout the discussions of this text, but certainly the point of origin of this converter topology is an arguable one.

When multiple outputs and/or DC isolation from input to output are desired in a buck-boost converter, it is not necessary to incorporate a DC transformer, as we will see. Consider the buck-boost circuit of Fig. 7.5A, where the inductor L now has two identical windings in parallel. This winding addition does not alter the basic converter operation. Next, the links connecting the two inductor windings are removed, as shown in Fig. 7.5A. Again, circuit operation remains the same except that DC isolation between the input and the output has been achieved. The final step in this process is shown in Fig. 7.5C, where additional secondary windings have been introduced in the inductor assembly with arbitrary turns ratios relative to the inductor "primary." This transformer-isolated multiple-output form of the buck-boost SPC is called the *flyback converter* and is probably the most popular circuit in use today for output power levels below 50 W. Its primary advantage is simplicity, particularly in that only one magnetic component is needed, even when input-to-output DC isolation and/or multiple outputs are required from the converter. The primary difficulty associated with this SPC is the design of inductor L, which now assumes the roles of both an energy-storage device and a transformer element. Its design becomes even more difficult when multiple outputs with good cross regulation are required. Successful flyback SPC design examples with as many as 25 isolated outputs do exist, but their magnetics designs were very complicated.

Besides the pulsating input and output current characteristic and the high switch currents mentioned earlier, the control-to-output small-signal transfer function of the buck-boost has a *right* half-plane zero, as we will see in Chapter 10. The presence of this troublesome zero should come as no surprise since one of the elementary constituents of the buck-boost converter has a similar control-to-output transfer function.

When a buck SPC is preceded by a boost SPC, as shown in Fig. 7.6A, quite a different, and in many ways a more useful converter can be derived in much the same manner as was followed to evolve the buck-boost and flyback SPCs. In Fig. 7.6B, the two switch-and-diode sets are replaced by a single DPDT switch. In this case, C1 cannot be omitted since it is required for energy storage. As was

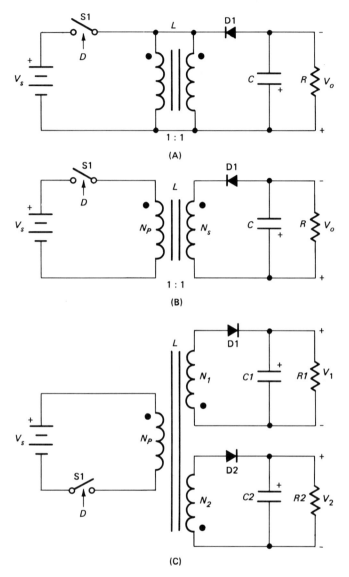

Fig. 7.5. Derivation of the flyback converter from the buck-boost converter.

the case for the buck-boost, if a polarity inversion of output voltage is permitted, the circuit can be further simplified. As indicated in Fig. 7.6C, the DPDT switch and shunt capacitor network can be replaced by a SPDT switch and a series capacitor. A practical realization of this new circuit is shown in Fig. 7.6D. This

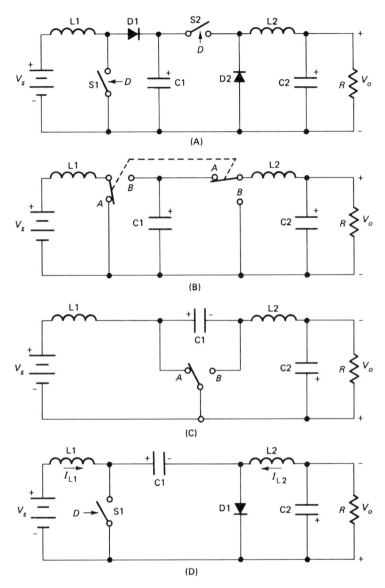

Fig. 7.6. Evolution of the Cuk converter.

SPC is known as the Ćuk (pronounced "Chook") converter after its inventor, Dr. Slobodan Ćuk of the California Institute of Technology [Ref. 6].

From a historical point of view, it is interesting to note that many popular SPC circuits were conceived more or less at random, and often the same SPC topology was invented by many different people at different points in time. SPC invention processes historically have been anything but orderly. The Ćuk converter is very much an exception to this development tradition. The topology was the result of a deliberate and logical effort by Dr. Ćuk to synthesize an SPC circuit with as many desirable conversion properties as possible. To the best of our present knowledge, *no* other independent invention of the basic Ćuk SPC topology has yet been claimed, although patents held by other power electronics engineers do exist which add further enhancements to the Ćuk topology.

Looking at the basic Ćuk SPC of Fig. 7.6D, we see that it requires only a *single* switch and a *single* commutation diode, with C1 serving as the main energy transfer link between input and output. When operating in the continuous mode, the input *and* the output currents of the Ćuk SPC are non-pulsating and associated AC ripple currents can be made arbitrarily small by increasing the values of inductors L1 and L2. This particular converter property often eliminates the need for additional input/output EMI filters in an application. Recall that both the boost and buck SPCs frequently require an additional output and input low-pass filter, as shown in Fig. 7.7A and 7.7B, respectively, to minimize conducted noise from SPC switching actions. Comparing the number of circuit components in the SPCs of Figs. 7.7A and 7.7B to that of the Ćuk converter in Fig. 7.7C, one finds that all three use the same number.

The Ćuk SPC, like the buck-boost converter, can provide an output voltage magnitude either above or below the value of its source voltage, depending on the duty cycle of S1 in Fig. 7.7C. Ideally the relationship between input and output voltages is simply $D/(1 - D)$. Switch, diode, and capacitor currents are comparable in the Ćuk converter to those of their counterparts in the buck-boost SPC operating with the same voltage transformation ratio and input power level. If input and output EMI filters are added to the buck-boost converter, the resultant conversion system becomes more complex than the Ćuk approach.

Like the cascaded buck SPC of Fig. 7.2A, the basic Ćuk converter has two inductors. It is readily possible to operate the converter so that one inductor is operating in the continuous current mode and the other in a discontinuous mode. This particular possibility is usually avoided in practice, since it leads to rather peculiar power conversion transfer characteristics!

DC isolation and multiple outputs can be added to a basic Ćuk converter by insertion of an AC transformer within its topology, as shown in Fig. 7.8. Here, the energy transfer capacitor (C1 of Fig. 7.7C) is divided into two series capacitors, C1A and C1B, whose total series capacitance value is equal to that of C1. A large inductor, such as that represented by the magnetization inductance of a transformer's primary, can be placed from the junction of C1A and C1B to the return point of source voltage *without* significantly affecting the converter operation. With the addition of secondary windings on this "inductor" along with associated output filter circuitry on each winding, the multi-output transformer-

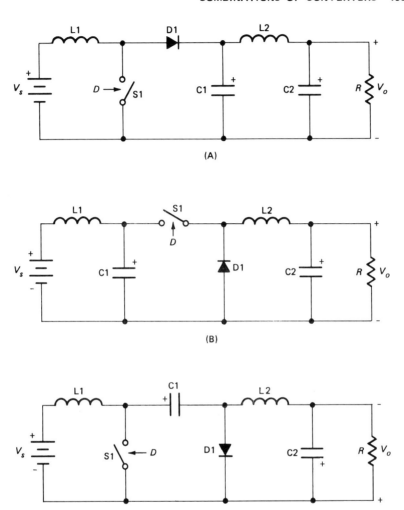

Fig. 7.7. Circuit comparisons of the boost, buck, and Čuk converters.

isolated Čuk converter results, as illustrated in Fig. 7.8B for a two-output case. Note that T1 does not store energy as part of the converter action. Because of the presence of capacitors in series with each winding, no DC can appear across the transformer windings. This allows full utilization of core flux capability. For these reasons, T1 is much easier to design, has better cross-regulation properties between outputs, and is usually smaller in physical size than the transformer of a flyback converter.

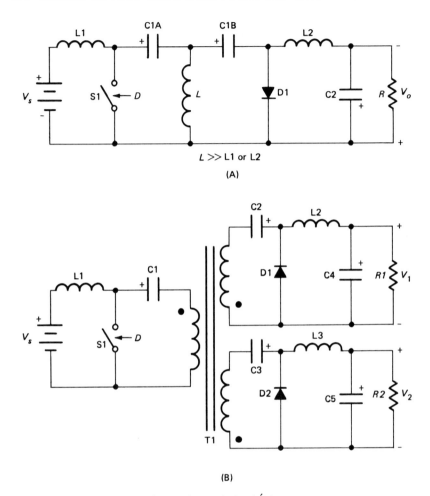

Fig. 7.8. The transformer isolated Ćuk converter.

Another unique enhancement can be added to the Ćuk converter, if desired. It is possible to have mutual inductive coupling (k) between L1 and L2, as shown in Fig. 7.9A. This can be achieved by placing L1 and L2 on a *common* core and controlling the value of k to reduce either input or output current ripple magnitudes. In fact, for properly selected values of coupling and turns ratios between the two inductor windings, input *or* output current ripple can actually be reduced to *zero*! This unusual effect is a function of the value of leakage inductances between the windings on L. However, in a practical sense, it is very difficult to build multi-winding inductances with closely controlled mutual-coupling properties, especially if the leakage inductances must be quite small.

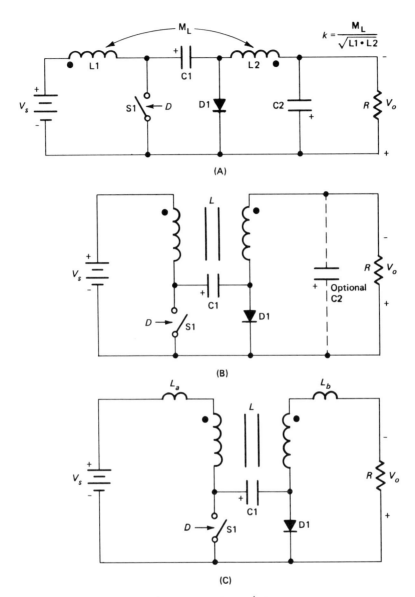

Fig. 7.9. The coupled inductor Ćuk converter.

So, as a practical expedient, L is usually built with tight coupling ($k \approx 1$) between windings and a small trimming inductor added on either the primary or secondary for current ripple reduction, as shown in Fig. 7.9C.

An interesting single-inductor version of the Ćuk converter is shown in Fig.

7.10B. In Fig. 7.10A, L has been moved from its original position above S1, D1, and C1 in Fig. 7.9B to one below these circuit elements. By making the turns ratio of the inductor in Fig. 7.10A one-to-one it can be reduced to a single-winding inductor as shown in Fig. 7.10C. This later circuit topology is simply a buck-boost circuit with its energy storage capacitor C1 connected directly to the positive side of the input source rather than to the return side.

The Ćuk converter has a multitude of variations, a few of which are shown in Fig. 7.11. Some of these unusual variations allow *both* input and output ripple current magnitudes to be made zero *simultaneously* using the magnetic coupling methods discussed earlier.

The process of cascading buck and boost SPCs can be carried another step farther by cascading a boost and a buck-boost converter, or a buck-boost converter, or a buck-boost converter and a buck converter. The results of performing the same network simplifications developed earlier on these combinations are shown in Figs. 7.12 and 7.13, respectively.

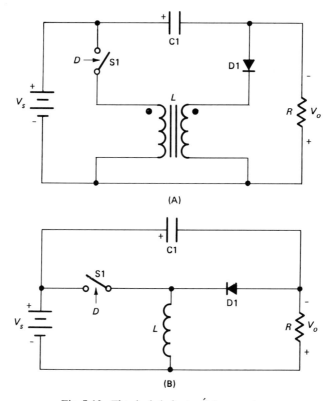

(A)

(B)

Fig. 7.10. The single inductor Ćuk converter.

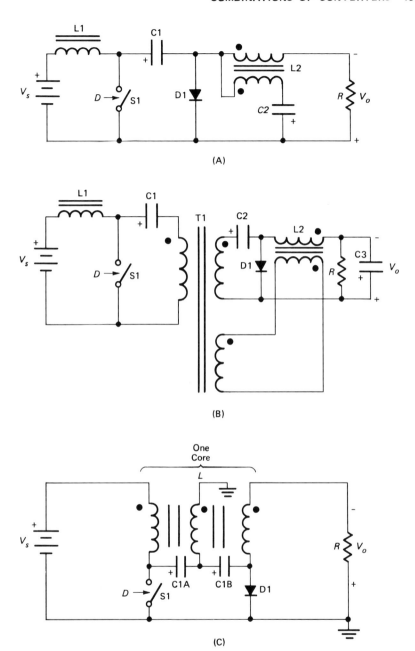

Fig. 7.11. Ćuk converter circuit variations.

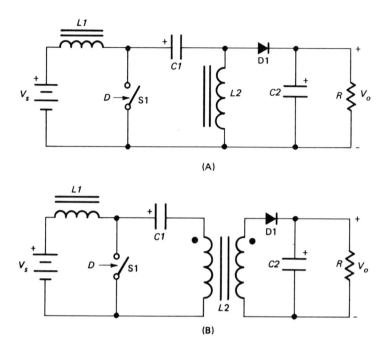

Fig. 7.12. Cascaded boost and buck-boost converters.

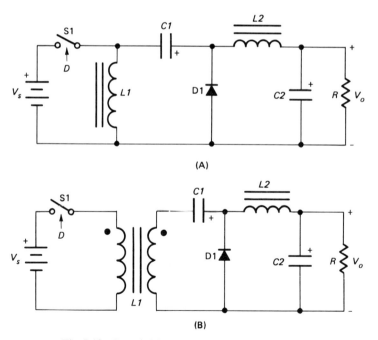

Fig. 7.13. Cascaded buck-boost and buck converters.

Even though we have essentially cascaded three converters in the last examples, the final circuit form of the simplified SPCs contain only a *single* switch and a *single* commutating diode. Based on these experiences, it appears that any SPC system formed by a number of cascaded, alternating buck and boost converters can be reduced to a topology consisting of one switch and one diode. From a practical standpoint, the need for more than two stages in cascade is rare.

7.3 PARALLELING CONVERTERS

There are a number of reasons for paralleling two or more converters, the most common being the need to provide higher output power levels. Given the practical limitations of circuit components (in particular, semiconductor switches), there will always be a maximum power output for any given converter design. Of course, it is possible to parallel multiple switches and commutating diodes to extend the output power capabilities of a single converter. However, this process usually involves derating individual device capabilities since exact sharing of current or voltage stresses is usually not achievable in a practical sense. Achievable power output from a single SPC with paralleled circuit elements may well be less than that possible from a conversion system composed of paralleled SPCs using single circuit elements. Unfortunately, there is no simple means for deciding which conversion system design approach will be the best for a particular output power need, other than to generate a fairly detailed design of each approach and then compare the results. At the present time, SPCs with power output levels above 10 KW usually use two or more converters operating in a parallel fashion.

Another reason for paralleling converters is higher system reliability as a result of operational redundancy. For example, if three converters are paralleled, any two of which can supply the full load power, the probability of two failures occuring in two of them that would disable the entire conversion system is much less than that of single failure in a conversion system composed of only one converter. Systems using five converters in which any three can fulfill full output load demands are not uncommon. Although complex in circuitry, very high functional reliability can be achieved by well-designed paralleled SPCs in a high power application.

Converters are often paralleled to increase the frequency and lower the magnitude of the input and/or output current pulsations. Potentially, this allows a reduction in the size and the weight of input and/or output EMI filter networks. An example of ripple current reduction by paralleling is shown in Fig. 7.14. Here, the switches (S1 through S3) of three paralleled flyback converters are operated sequentially, one-third of a complete switching cycle apart from one another. This timing sequence results in input and output current ripple frequencies three times that of the overall converter switching frequency. Another

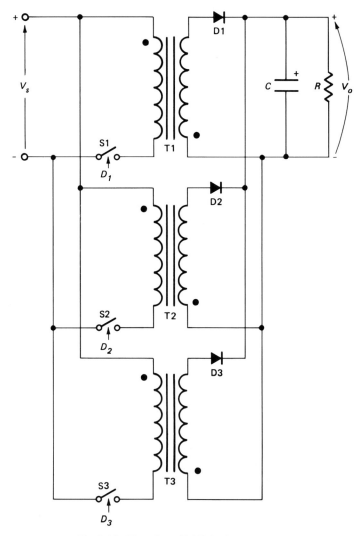

Fig. 7.14. Phased parallel flyback converters.

benefit of sequenced switching is an increase in the duty cycle of input/output current waveforms and a decrease in their peak currents. For example, in Fig. 7.14, for $D_1 = D_2 = D_3 = 0.3$, the input current duty cycle will be 0.9 with output current duty cycle unity, since the output current pulses from each converter will overlap. This overlap results in a nearly continuous output current waveform. Also, the peak value of the input current will be one-third that of a single flyback converter of comparable output power level.

Production power supplies using SPCs must meet the requirements of widely varying applications and still be low in cost. A common product approach adapted by many power supply companies is to design a single medium-power converter and to produce this single design in quantity. Customer requirements for higher power outputs than possible from the single converter are met by paralleling two or more of the basic designs and adding necessary auxiliary control electronics. This product design philosophy results in low nonrecurring engineering costs as well as lower recurring manufacturing expenses.

There is one problem which is common to all parallel converter schemes, namely how to share the output load demands equally among each SPC. Other than potential similarities in circuit and component impedances, there is no inherent mechanism to force current-sharing between paralleled converters. Therefore, large imbalances in load sharing can occur if several converters are simply wired in a parallel fashion. The usual solution to this problem is the use of regulatory current sensors to maintain balance in load sharing among the converters. Control schemes which compare individual switch currents against an error voltage related to total output load are viable methods to insure current sharing among SPCs.

7.4 OTHER SPC COMBINATIONS

As shown in Figs. 7.1C and D, the inputs and/or the outputs of converters can be connected in a series fashion. An example using two parallel quasi-squarewave converters with their inputs tied in parallel and their outputs connected in series is shown in Fig. 7.15.

If the required output voltage is high enough that the additional voltage drop of one more rectifier diode can be tolerated, the SPC system of Fig. 6.15 has some definite advantages. Because both transformer output windings are connected in series, their currents must be the same. This current relationship is reflected to transformer primary circuitry so as to force equal input currents in each converter. This automatic power-sharing feature eliminates any need to incorporate auxiliary electronics to insure balance. The output diodes of Fig. 7.15 will also see equal and lower reverse voltage stresses than a single SPC of this type. Here, again, equal voltage stresses on these diodes are results of the current-sharing feature mentioned earlier.

The advantages of automatic sharing of reverse diode voltages could be gained by using series multiple connections of windings on a single transformer. But, for very high output voltages on the order of 20 kV or greater, a single transformer with multiple windings for voltage-sharing can be very difficult to build. On occasions, it may be simpler to use separate transformers for each winding and increase the insulation on windings in the string that see higher voltages.

Input power sharing can be obtained by connecting the inputs of SPCs in series.

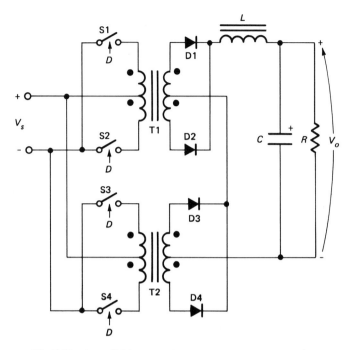

Fig. 7.15. A parallel-input, series-output converter connection.

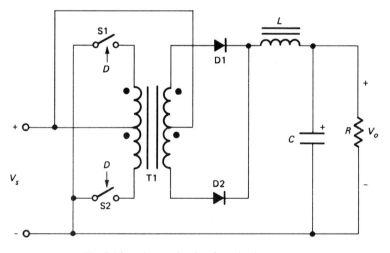

Fig. 7.16. A boost circuit using a buck converter.

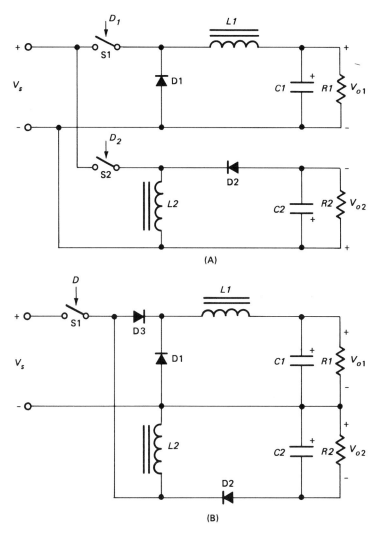

Fig. 7.17. A scheme for combining duty cycle and frequency modulation to provide two regulated outputs with one switch.

Again, automatic input current balance is provided and applications with very high input source potentials can be met with little difficulty by an input series arrangement of two or more SPCs with lower input voltage capability.

In the converter of Fig. 7.15, the voltage waveform present at the input of the output low-pass filter is the algebraic sum of the potentials of the individual transformer windings. Therefore, it is quite possible with this SPC to synthesize a "stepped" voltage waveform with a predetermined harmonic content. This can be accomplished by using different winding turns ratios in T1 and T2 and by phasing switch conduction duty cycles between the two converters. The number of voltage "steps" can be increased by connecting more converters in this series fashion—properly phased, of course. This phasing method can be used to reduce the size of the output low-pass filter network or to synthesize an output waveform that closely approximates a sinusoid in its shape.

Many other SPC combinations are possible and are occasionally seen in practice. For example, if the desired SPC output voltage magnitude is somewhat greater than the highest value of input voltage and DC isolation is not required, then the converter shown in Fig. 7.16 can be used. In this SPC, the input voltage is in series with the converter's output. The duty cycle controls for the primary switches are then set to allow most of the input power to the system to "bypass" the converter when the input voltage is at its maximum. As input voltage is reduced, more power is processed by the SPC. Since only a small portion of the total output power is controlled, the size and cost of the converter is reduced while improving overall efficiency of power conversion. This converter performs a voltage-boost function, even though it is a buck-derived SPC.

A very unusual SPC system is shown in Fig. 7.17, which has recently been reduced to practice. The object of the converter in Fig. 7.17A is to provide two separate, well regulated output voltages. However, one converter switch and any associated drive networks can be eliminated, as shown in Fig. 7.17B, if one converter is operated in the continuous mode and the other in the discontinuous mode. The output voltage of the converter operating in the continuous mode is determined by D and is relatively insensitive to variations in switching frequency. However, the output voltage of the converter operating in the discontinuous mode is a function of both D and switching frequency.

By using a variable duty cycle control to regulate the continuous mode output and a variable-frequency control to regulate the discontinuous mode output, simultaneous regulation of both outputs can be achieved. Note in Fig. 7.17B that a blocking diode (D3) is required to prevent the two converter modes from interacting with one another.

Obviously we can continue the combinational exercises of this chapter indefinitely. The examples presented, however, should serve to illustrate the many possibilities and utility of converter aggregation processes.

8. Magnetic Component Tapping

Rarely do the needs of a particular power processing application require that a new SPC be created, especially since many variations are already known, as we have just seen in the last two chapters.

However, occasions do arise when available SPC topologies are almost but not quite good enough for an application and some small changes in their electrical characteristics are necessary. One of the simplest means to enhance performance is the tapping of the windings of the inductor and transformer elements. Many beneficial results can be realized and include circuit simplification, improved cross-regulation properties, lower EMI noise levels, component stress reductions and, in some special cases, the elimination of moving left half-plane poles as well as right half-plane zeros in the small-signal transfer functions. Most converter circuits are amenable to magnetic tapping for specific improvements such as those just described.

We will begin this chapter by discussing inductor-tapping methods in SPCs, followed by examining related tapping of any associated transformer windings. Finally, techniques will be covered whereby combinations of inductor and transformer tapping methods can be useful in SPCs.

8.1. INDUCTOR TAPPING

One of the simplest and most common examples of inductor tapping is shown in Fig. 8.1A, where the switch in a boost converter is connected to a tap on the winding of the inductor.

A brief comparison between important tapped and untapped boost converter electrical characteristics is given in Table 8.1. Ideal components are assumed and the converters are both operating in the continuous mode. We see that as the switch is moved to a tap on the inductor, switch OFF voltage stress is reduced, (remember V_o is always greater than V_s) which allows one to have a higher output voltage for a given switch breakdown rating. However, there is a price to pay for this improvement. The diode reverse voltage increased over that seen in the untapped converter and, as shown in Fig. 8.1B, the input current will now have

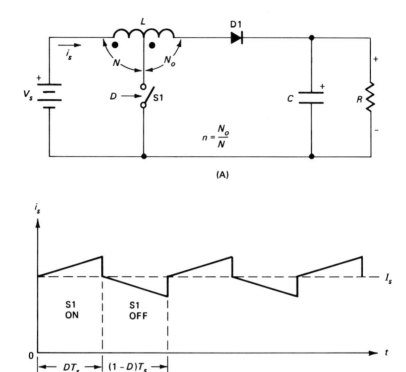

Fig. 8.1 A tapped-inductor boost converter.

discontinuities in its waveshape. Since power diodes with large reverse voltage ratings are usually much easier and cheaper to obtain than switches with high OFF voltage capabilities, a reduction in switch voltage stress by tapping can certainly be advantageous.

From Table 8.1, the value of n in the tapped inductor converter will always be greater than zero; therefore:

$$M = \frac{1 + Dn}{1 - D}.$$ (8.1)

In the case of the converter with an untapped inductor:

$$M = \frac{1}{1 - D}.$$ (8.2)

TABLE 8.1. Boost Converter Comparisons—
Continuous Mode.

Parameter	Untapped	Tapped
V_T	V_O	$(V_O + nV_s)/(n + 1)$
V_{DR}	V_O	$V_O + nV_s$
D	$(M - 1)/M$	$(M - 1)/(M + n)$

V_T = peak switch off voltage
V_{DR} = diode reverse voltage
M = ideal voltage transfer factor
n = tap factor (≥ 0)

From Eqs. (8.1) and (8.2), it is apparent that, to achieve a given value of M in both converters, a *smaller* value of D will be needed in the case of the tapped inductor SPC. This reduction factor is equal to $1/(1 + (n/M))$.

In a similar fashion, the diode of the boost SPC can be connected to a tap on its inductor as shown in Fig. 8.2A. In contrast to the previous tapping method, this movement reduces diode reverse voltage at the expense of increasing the OFF voltage of the switch. In the buck converter, inductor tapping as shown in Fig. 8.2B and C can also be used to alter the component stresses and voltage gains.

During the time intervals when all primary switches in a quasi-squarewave converter (Fig. 5.13) are OFF, the current in the primary windings will ideally be zero. However, the inductor current in the secondary must commutate through D1 (if present) and/or through diodes D2 and D3 during such intervals. To maintain ampere-turn balance in the secondary winding of T1, the commutating current must split equally in each half of this winding (assuming D1 is not present). Because there will be small differences in the conduction characteristics of diodes D2 and D3 as well as in secondary winding impedances, it is very probable that the average currents (I_1 and I_2 in Fig. 8.3A) will not be equal. As a result, an unbalanced DC bias on the transformer core will occur and can cause core saturation. This particular problem is especially severe if ungapped tape cores are used in the transformer. The addition of a third commutation diode (D1) does not guarantee that saturation will not occur, as it only provides another potentially imbalanced impedance path for a portion of the inductor current. However, if D2 and D3 are connected to a tap on the output inductor as shown in Fig. 8.3B, then both secondary diodes can be reverse-biased during the intervals when the primary switches are OFF. Diode D1 now must commutate the entire inductor current, and therefore, no DC core bias can occur as a result of imbalance in rectifier diode impedances. Reverse-bias voltage levels for D2 and D3 need only be 0.5 to 1.0 V, so that the tap ratio can be very small. There will be a discontinuity in the inductor current due to the tap. However, except for

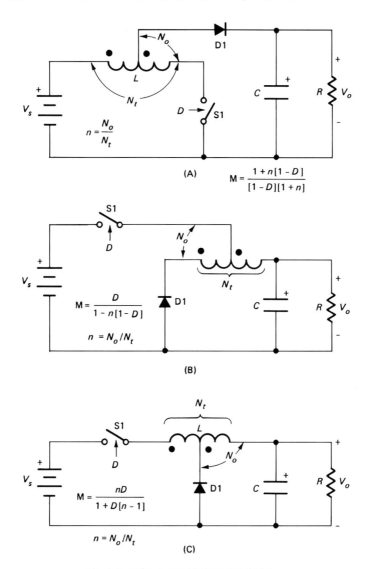

(A) $$M = \frac{1 + n[1 - D]}{[1 - D][1 + n]}$$

$n = \dfrac{N_o}{N_t}$

(B)

$$M = \frac{D}{1 - n[1 - D]}$$

$n = N_o/N_t$

(C)

$$M = \frac{nD}{1 + D[n - 1]}$$

$n = N_o/N_t$

Fig. 8.2. Other tapped-inductor variations.

very low output voltages, this discontinuity will be quite small with little contribution to the rms value of the inductor current.

Although transformer core imbalance from the effects of secondary commutating currents is eliminated by the inductor-tapping technique just described, a potential problem could arise from its use. With diodes D2 and D3 in Fig. 8.3 now reverse-biased at times when the SPC's primary switch networks are OFF,

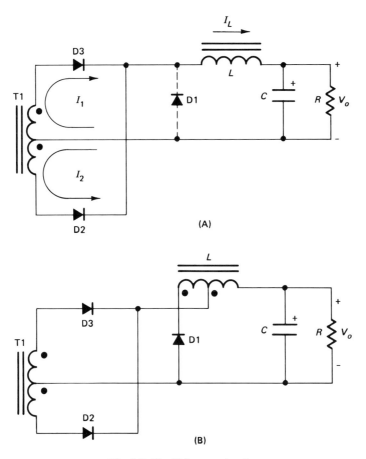

Fig. 8.3 The McLyman circuit.

magnetization energy within T1 is still present and must be eliminated. This is usually accomplished by adding dissipative energy clamps across primary switch elements. If this energy is large, overall SPC system efficiency can suffer from its loss in protective circuitry.

The concept of tapping can be extended to the use of multiple windings on the inductor. Recall that one of the disadvantages of a multi-output quasi-squarewave converter is the need for a separate energy storage inductance for each output. Also, if reasonable cross regulation between outputs is to be maintained, each output inductor must remain in the continuous mode. Both of these disadvantages can be circumvented by using a *single* multiwinding inductor assembly as shown in Fig. 8.4. Here, the turns ratios of the inductor's windings are chosen to be the same as the ratios of corresponding transformer windings.

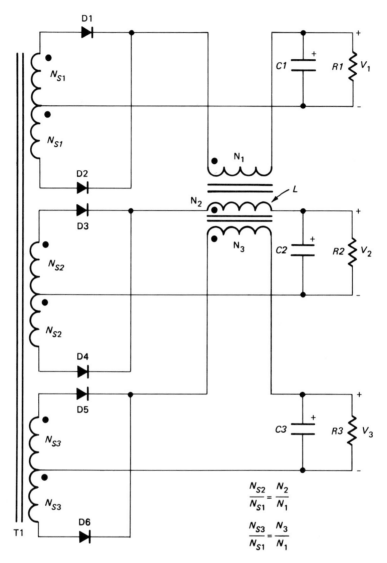

Fig. 8.4. A multiple-output quasi-squarewave converter using a single multi-winding inductor.

The advantages of this common arrangement are several. First, the inductor operating mode is determined by the sum of the ampere-turns of all its windings. Therefore, the inductor can remain in the continuous mode even if one or more output currents fall to a small value. This allows reasonable regulation of all SPC outputs regardless of load. Since each low-pass filter does not need to have a separate inductor, a significant reduction in number magnetics is achieved. Also, commonality of core permits less total energy to be stored in the inductor to maintain continuous conduction, thus reducing the physical size of L. Even if the total stored energy were the same, the weight of the multiwinding inductor assembly would still be less than the sum of weight of separate inductors.

The design of the multiwinding inductor is a straightforward procedure. We know that for a single output to maintain continuous conduction mode:

$$L_1 \geqslant \frac{R_{L1}(1-D) T_s}{2} \tag{8.3}$$

where T_s is total converter switching period, R_{L1} is the minimum load expected, and D is the conduction duty cycle of the SPC primary switches. In the multi-output case, R_{L1} is the sum *total* of all output loads transformed to a single output for reference. For example, if we chose the first output of the SPC in Fig. 8.4 as our reference, then:

$$R_{L1} = \frac{V_1^2}{P_1 + P_2 + P_3} = \frac{V_1^2}{P_T} \tag{8.4}$$

where P_1 through P_3 are the individual minimum output power levels. The inductance associated with the first secondary is then calculated from Eq. (8.3) as:

$$L_1 \geqslant \left[\frac{V_1^2}{P_T}\right] \left[\frac{1-D}{2}\right] T_s. \tag{8.5}$$

Allowing for the rectifier diode forward conduction voltage (V_o) and resistive voltage drop in secondary windings (V_w), the turns required for the other windings must be:

$$N_n = \left[\frac{V_n + V_{Dn} + V_{wn}}{V_1 + V_{D1} + V_{w1}}\right] N_1, \quad n = 2, 3. \tag{8.6}$$

With L_1 and the turns ratios between windings known, the inductor design can proceed in a normal fashion. One word of caution is needed here—maintaining an equality of balance in inductor turns ratios to corresponding transformer secondary values is essential to prevent "circulating" currents in output filter networks. Severe unbalance can produce high ripple currents and potential damage to output filter capacitors.

A multiwinding or tapped inductor can also be used to alter the ripple current in a filter capacitor (C) as shown in Fig. 8.5A. The average DC voltage appearing on the capacitor will remain the same. The use of the auxiliary winding (or tap)

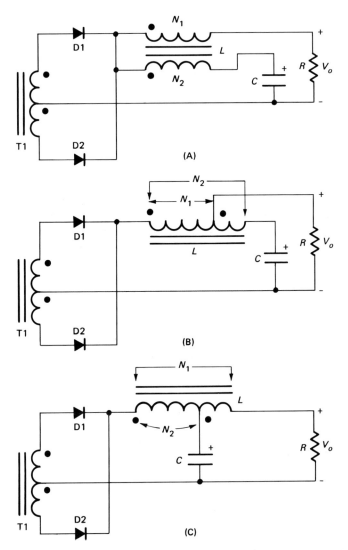

Fig. 8.5. Schemes for increasing the effective value of the output filter capacitor.

allows a selection of C that often results in a smaller and less expensive part. If the impedance of the filter capacitor at high frequencies is dominated by internal parasitics (resistive or inductive), then the scheme of Fig. 8.5A may not be advantageous. The companion circuit of Fig. 8.5B is equivalent to that of Fig. 8.5A when $N_2 > N_1$, while the circuit in Fig. 8.5C is equivalent to that of Fig. 8.5 A when $N_2 < N_1$.

Further discussions of integrating inductor as well as transformer windings on single core structures are contained in Chapter 12.

8.2. TRANSFORMER WINDING TAPS

Provision for a multiplicity of output voltages is a very common requirement for an SPC. Traditionally, this requirement is met by providing multiple transformer windings, as shown in the example of Fig. 8.6A. Normally, one output is chosen to be directly regulated and voltage regulation of the other winding outputs may suffer as a result. One of the key ingredients that determine cross-regulation properties between SPC windings is winding-to-winding coupling. If windings are tightly coupled, then good cross regulation is possible. Coupling between windings can be greatly improved by using a single secondary winding, tapped as shown in Fig. 8.6B. However, DC isolation between windings is lost using this technique. In addition to improving the winding coupling properties, copper utilization factors are usually better. Voltage drops due to winding resistance are somewhat self compensating because the resistance is common to both windings, which further improves the cross regulation. Another factor determining cross regulation is the variation of the forward voltage drops (V_f) of the rectifier diodes. For silicon diodes, the temperature coefficient of V_f is about -2 mV/°C. In applications where extreme changes in circuit temperatures are present, definite changes in V_f will occur and cross regulation will suffer as a result.

For example, assume that one output (V_2) is +5 V and the other (V_1) is +10 V in the circuit of Fig. 8.6B. Further, we will assume that the +5 V output is tightly regulated by the control loop. If the forward voltage drops of D2 and D3 were to increase by 50 mV resulting from a 25°C drop in temperature, the voltage control electronics would respond by increasing the winding voltages associated with D2 and D3 by 50 mV to maintain a constant +5 V output. Since the ratio of N_1 to N_2 is 10/5, the winding voltages at D1 and D4 will increase by (10/5) (50 mV), or 100 mV. Presumably, the voltage drops of D1 and D4 will also increase by 50 mV due to the temperature change. Therefore, a net 50 mV increase of the +10 V winding will occur. In this example, it is possible to compensate for the +10 V output change by adding a series diode (D5 in Fig. 8.6B). The presence of D5 adds further power loss to the SPC and could reduce efficiency significantly if loads on the +10 V output are large.

The primary windings of an SPC's transformer can also be tapped to enhance performance. Fig. 8.7A shows a quasi-squarewave converter (Fig. 5.13), in which two more switch elements (S3, S4) with corresponding taps on the transformer have been added. Note also that each of the two new switches are operated at a 50% duty cycle. The result of this change in topology is an alteration of the input current and output voltage waveforms as shown in Fig. 8.8. For illustration purposes, we have assumed an average output voltage of 5 VDC and an average

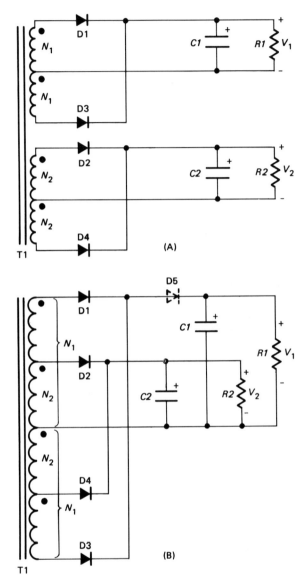

Fig. 8.6. Multiple-output secondary winding variations.

input current magnitude of 2 ADC. Fig. 8.8A and C are the waveforms for an "untapped" converter (Fig. 5.13) and Figs. 8.8B and D are corresponding waveforms for the tapped converter of Fig. 8.7A. Note that the primary-to-secondary turns ratios have been adjusted so that D is the same in both converters. By

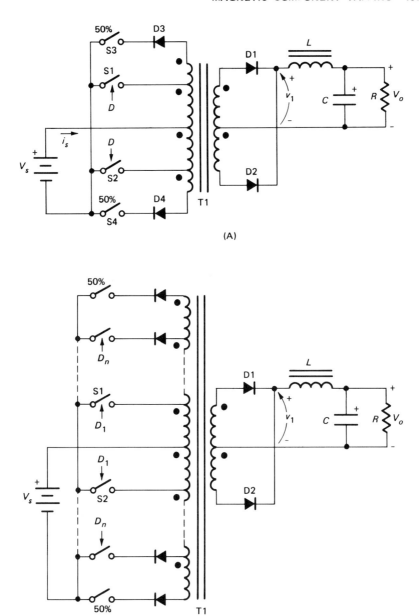

(A)

(B)

Fig. 8.7. A tapped-primary quasi-squarewave converter.

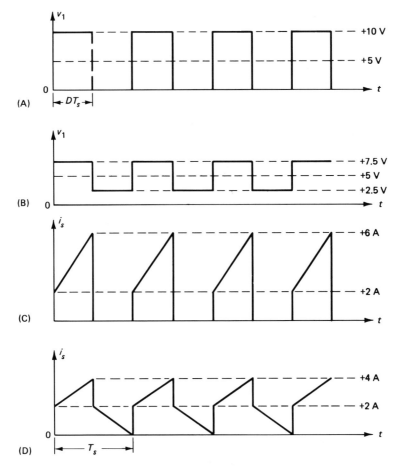

Fig. 8.8. Voltage and current waveform comparisons.

inspection of the waveforms of Fig. 8.8, several benefits become evident. First, the rms value of v_1 is lower in the tapped case, which will result in the output filter being smaller in size. Also, the peak reverse voltage on output rectifier diodes is reduced. In the tapped example, the rms value of input ripple current as well as peak switch current is smaller, resulting in a smaller input filter network.

However, these improvements are not without penalty. The converter of Fig. 8.7A is now more complex, with twice as many primary switches and a multi-tap transformer primary winding. In the standard quasi-squarewave converter of Fig. 5.13, a lower limit on input voltage value is reached where its primary switches are operating at their maximum duty cycle. Also, there is no inherent upper limit on input voltage value other than the stress limitations of the circuit components. By reducing the switch duty cycle, a very wide range of input voltages can be accommodated by the conventional quasi-squarewave SPC. In the tapped version of this SPC in Fig. 8.7A, there will be an inherent upper limit on input voltage with its magnitude determined by transformer tap locations. To realize the benefits as outlined above for this converter, the range of input voltages is usually constrained to 2 to 1 or less.

Multiple transformer taps can be used, as shown in Fig. 8.7B, to synthesize an output voltage (v_1) with as many "steps" in level as there are taps. This approach is sometimes used by SPC designers to generate outputs that closely approximate sinusoids. Primary diodes in series with all but the innermost switches are necessary since the voltage appearing across the outside switches will reverse polarity when an inside switch is closed. Therefore, these diodes prevent any reverse tap currents which might damage switch components or produce converter malfunction.

The idea of using primary taps has many variations, some of which are shown in Fig. 8.9. Here, the number of switches or diodes can be reduced by altering the topology slightly (e.g., see Figs. 8.9A and C) or by moving the return reference point (Figs. 8.9B and D).

Primary transformer taps with additional switch elements can also be used in other circuits. For example, Fig. 8.10 shows a tapped version of the boost-derived converter of Fig. 6.4B. Bridge versions of the converter in Fig. 8.7A are shown in Figs. 8.11 and 8.12. If the input voltage range is 2 to 1 or less, then only one additional switch is needed, as illustrated in Fig. 8.11B. When using a full-bridge primary circuit, there are several alternate ways in which transformer taps and switch elements can be interconnected. Two of these possibilities are shown in Fig. 8.12.

In a similar fashion, primary transformer tapping can be applied to half-bridge SPC circuits. One example is shown in Fig. 8.13.

Secondary-tapping with auxiliary switch networks is also possible. Fig. 8.14A shows a tapped secondary version of the post-regulated buck-derived converter developed in Chapter 5 (Fig. 5.14A). This SPC has exactly the same electrical properties as the tapped primary quasi-squarewave circuit of Fig. 8.7A, except that voltage regulation control (S3) is performed on the secondary and three switches are needed for a single output rather than four in the circuit of Fig. 8.7A. If multiple regulated SPC outputs are desired, then each output will still need a separate control switch, as was the case in Fig. 5.14B.

A boost-derived circuit using a tapped secondary is shown in Fig. 8.14B. It's untapped counterpart was developed in Chapter 6 (Fig. 6.4B).

When an SPC's output rectifier network uses a bridge configuration, then several other tapped variations are possible, two of which are shown in Fig. 8.15.

Very high frequency or very high power SPCs frequently have very few turns on the transformer windings associated with low-voltage outputs. If multiple-

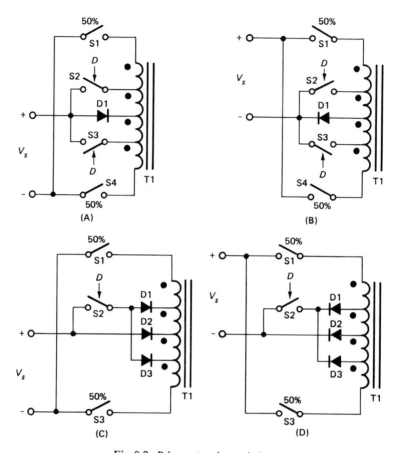

Fig. 8.9. Primary tapping variations.

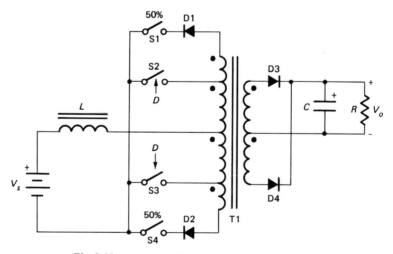

Fig. 8.10. A tapped primary boost-derived converter.

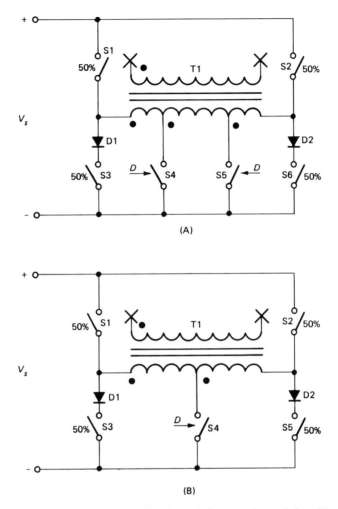

Fig. 8.11. Tapped primary full-bridge switch connection variations (I).

output voltages are required of such SPCs, a problem can exist in obtaining sufficient resolution and cross-regulation between all outputs since secondary volts-per-turn factors may be quite high. One means to alleviate this particular problem is the addition of a small transformer (T2) to provide a trimming voltage in series with the output rectifiers, as shown in Fig. 8.16. This trimming voltage may be either aiding or opposing in its contribution. Voltage amplitudes presented by these trimming windings are determined by the turns ratio of N_1 to N_2 and N_3. The primary turns (N_1) on T2 can be made arbitrarily large so as to provide as

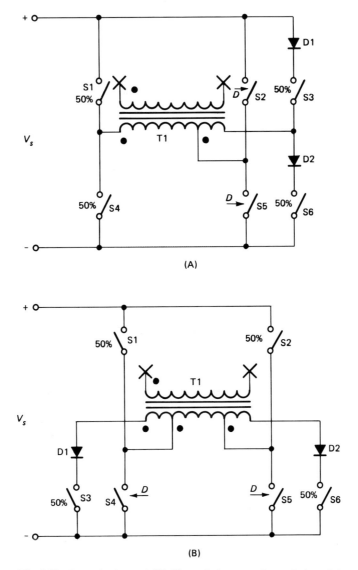

Fig. 8.12. Tapped primary full-bridge switch connection variations (II).

small a trimming voltage as desired. Therefore, corresponding resolution of output voltage (V_1) is very good. The addition of transformer T2 is not desirable for cost and size reasons; however it usually is a small transformer and relatively inexpensive. If improvements in resolution of several output windings are desired, transformer T2 can be designed to accommodate multiple trimming

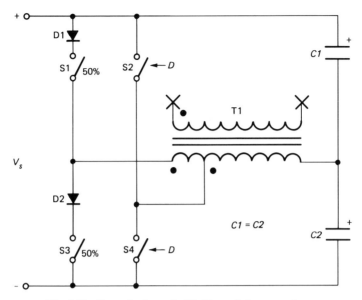

Fig. 8.13. Tapped primary half-bridge switch connection.

windings. Therefore, a separate "trimming" transformer will not be required
for each output in such cases.

Like taps, multiple transformer windings can be very useful in a DC-DC converter. For example, in the circuit of Fig. 5.8, DC-isolated drives are required
for all of primary switch elements. By using extra primary windings as shown in
Fig. 8.17, all switch drives can be referenced to the same return point. This can
permit (in many cases) the elimination of the drive transformers. Although the
design and construction of T1 will be slightly more complex, overall system
component count is reduced with corresponding savings in space and cost. This
particular technique can be applied to many of the SPCs discussed in earlier
chapters of this book to obvious advantage.

Another beneficial application of multiple primary windings in SPCs is shown
in Fig. 8.18. Here, two separate primary windings are used, each with its own
set of switches and one isolation diode. This arrangement allows the converter
to operate from either of two different power sources. As an application example,
V_{s1} could be a low voltage battery source and V_{s2} could be an AC-rectified source
of high voltage DC. The primary turns ratios of the transformer could then be
adjusted so that power is normally taken from V_{s2} until its voltage value drops
to:

$$[V_{s2} - V_{Ds2}] = \frac{N_2}{N_1} [V_{s1} - V_{Ds1}]. \qquad (8.7)$$

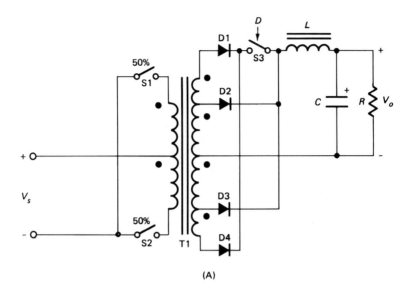

(A)

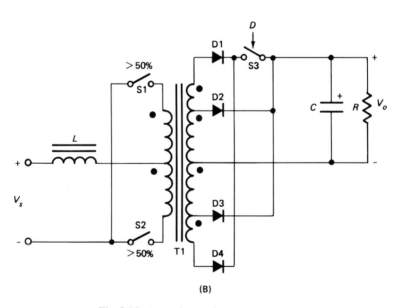

(B)

Fig. 8.14. Tapped secondary converters.

At this point, the converter's load power source is shifted to V_{s1} automatically with no interruption of output voltage during transition. As an additional refinement, a battery charger could be added across D_{s1} to charge V_{s1} during normal

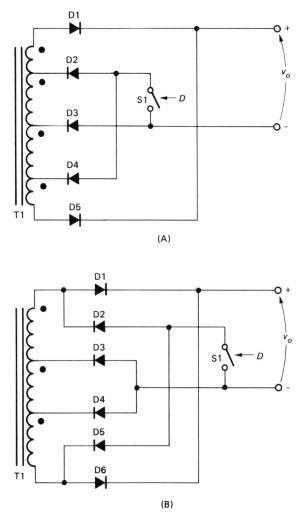

Fig. 8.15. Tapped secondary variations.

operation. Additional diodes (shown dotted in Fig. 8.18A) will be needed across S1 and S2 to provide a bidirectional path for charging current if switches S1 and S2 are *not* operated in phase with S3 and S4, respectively.

8.3. COMBINING TRANSFORMER AND INDUCTOR TAPPING METHODS

It is evident from earlier discussions that tapping of either the inductor or the transformer windings of an SPC can produce useful and beneficial results. Both

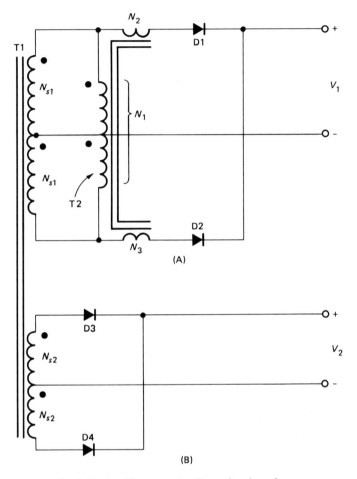

Fig. 8.16. Auxiliary output voltage-trimming scheme.

of these tapping techniques can be combined in a single converter so as to produce additional benefits. As a general rule, when both the transformer and the
inductor are tapped, the turns ratio of both taps will have the same value.

The SPC circuit of Fig. 5.6 in Chapter 5 was developed to provide high conversion efficiency when operating from a low voltage source of power. Conversion power losses due to conduction voltage drops in S2, S3, D1, and D4 of Fig.
5.6 can be reduced even further if both the inductor and the transformer primary
are tapped. The resultant topology is as shown in Fig. 8.19. In this circuit
modification, the average current flowing through primary diodes (D1, D2) is
reduced by the equal tap turns ratios of L and T1. Therefore, the overall ef-

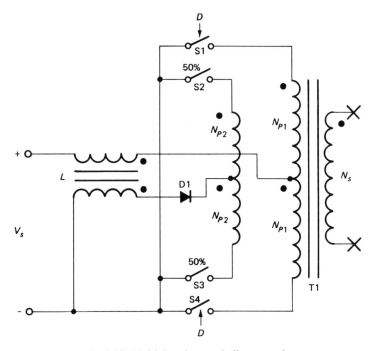

Fig. 8.17. Multiple primary winding example.

ficiency of power conversion will be much improved when a low voltage source is present.

In order for the output current to remain continuous, the turns ratios of the taps on the transformer and the inductor of a buck-derived SPC must be the same. If they are not equal, the inductor current waveform will have discontinuities with a corresponding increase in output voltage ripple. In addition, a small-signal analysis of an unbalanced tapped SPC of this kind will show that the cut-off frequency of the converter's equivalent output low-pass filter becomes a function of converter duty cycle, D. Therefore, without equal tap ratios, any buck-derived SPC will begin to display boost-like electrical properties.

The tapped transformer boost-derived SPC illustrated in Fig. 8.10A can have an alternate common connection point for its switches. This connection change is shown in Fig. 8.20A. The SPC's gain function is still the same even though its topology looks somewhat different. It is still a boost-derived converter with a discontinuous output current waveshape, moving low-pass filter poles, and a right-half-plane zero in its control-to-output transfer function. If the input inductor is tapped with the same turns ratio as the primary windings of the transformer, and two of its switches connected to the inductor tap as shown

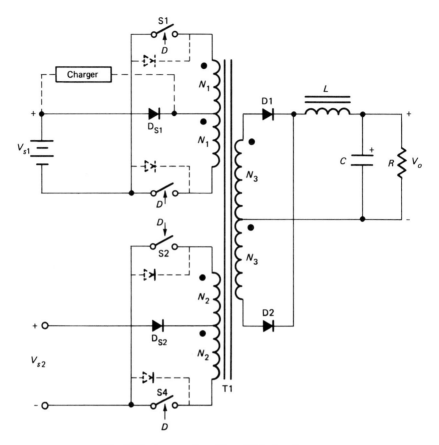

Fig. 8.18. A converter with multiple input sources.

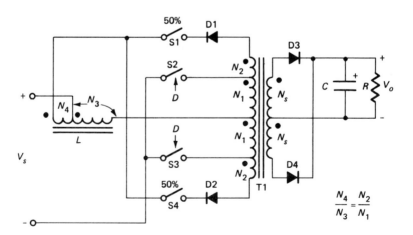

Fig. 8.19. A tapped inductor-and-transformer version of the Cronin converter.

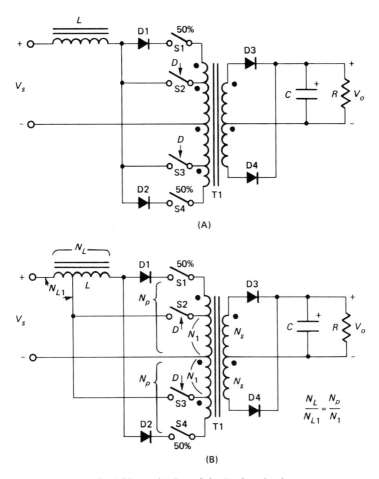

Fig. 8.20. Derivation of the Hughes circuit.

in Fig. 8.20B, a rather remarkable change in circuit properties occurs. The ampere-turns product in the primary winding of T1 is now constant and independent of switch conduction duty cycle, D. The output current waveform is now continuous, just as it would be for a buck-derived converter. Further, a small-signal analysis of the SPC topology will show that the effective value for output filter inductance is constant. As a result, the output filter poles do not move with duty cycle and the control-to-output right-half-plane zero, characteristic of a boost-derived converter, has disappeared! Despite the fact that this particular SPC circuit was clearly derived from a boost topology, it has many of the properties of a buck-derived converter. However, some properties characteristic of a boost SPC are still retained. For example, magnitudes of output

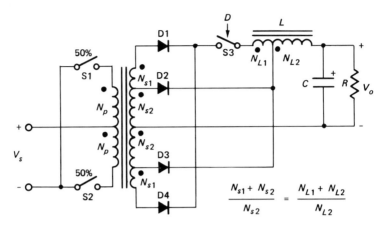

Fig. 8.21. A tapped transformer secondary and inductor circuit.

short-circuit and turnon inrush currents (see Chapter 3) cannot be controlled by control of switch duty cycles. Also, some circuit provisions must be made to safely dissipate the energy stored in the input inductor as well as in the transformer when all four switches in Fig. 8.20B are turned OFF.

In a similar fashion, the buck-derived converter of Fig. 8.14 can be equipped with tapped inductor and transformer elements as shown in Fig. 8.21. As was the case in the last example, SPC circuit properties are radically altered. If the transformer and the inductor are tapped equally, the input current of this SPC will be continuous just as it would be in a boost-derived converter. Output current is now discontinuous and the output filter transfer function will change as a function of the conduction duty cycle of switch S3. In contrast to the SPC of Fig. 8.20B, this converter topology has become boost-like in nature!

In this latest group of complex SPC circuits, it is interesting to observe that pulsating current characteristics of input or output terminals can be interchanged by adding or removing taps on internal inductors and transformers. From an EMI point of view, these design alternatives can be very useful in directing and controlling the major conduction noise sources of an SPC, especially if the converter must be designed to meet tough conducted EMI specifications such as those imposed by European countries on power supply systems.

9. Duality Among SPC Circuits

After examining a multitude of SPC topologies in preceding chapters, one might be led to believe at this point that all variation possibilities have been exhausted. Such is not the case.

Recall that the tools we used to build the more complex SPCs of past chapters were essentially combinational in nature. In this chapter, we will turn to evolutionary methods that employ deductive reasoning based on duality.

Duality is a fundamental electrical circuit principle and discussions of its application in linear network analysis are found in most basic electrical engineering textbooks. Although SPCs are *nonlinear* electrical systems by their very nature, we saw in Chapters 2 and 3 that we could formulate *linear* small-signal models for the buck and boost converters, using state-space averaging methods. Therefore, it is logical to assume that *linear* duality analysis and synthesis principles can be applied to SPCs.

In this chapter, we will conduct duality investigations of SPC topologies to generate new circuit variations and, perhaps more importantly, to show functional relationships between them that will enhance our understanding of their unique properties. As a part of these duality discussions, the concept of varying the SPC circuit properties by changing the switch operating sequences will be introduced.

9.1. BILATERAL INVERSION

In Chapter 4, it was demonstrated that bidirectional power transfer could be achieved in a DC transformer if switch elements were placed across the output diodes and antiparallel diodes placed across the primary switches (Fig. 4.16). This same technique can be used to provide bilateral power flow in any SPC, regardless of its complexity. An example of its application using a boost converter is shown in Fig. 9.1. Fig. 9.1A shows the elementary topology of the converter. In Fig. 9.1B, a switch (S2) is placed across the output diode (D1) along with another diode (D2) across the input switch (S1). The direction of power flow will now depend on switch conduction duty cycle (*D*) and the relative amplitudes

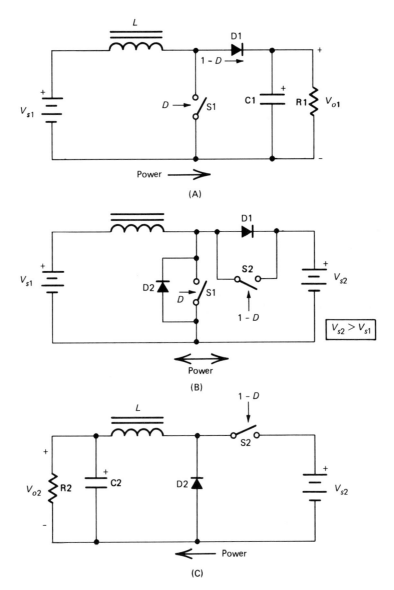

Fig. 9.1. An example of bilateral inversion.

of voltage sources V_{s1} and V_{s2}. If V_{s1} and V_{s2} are maintained relatively constant in their values, then the direction of power flow will be controlled by D. As D is made larger, the power will begin to flow from V_{s1} to V_{s2}. Conversely, as D is made smaller, direction of power flow will reverse and flow from V_{s2} to V_{s1}.

Continuing the bilateral process, the original switch (S1) and diode (D1) in Figs. 9.1A and B are now removed, resulting in the SPC of Fig. 9.1C. This converter circuit is now unidirectional in its power flow; however, in contrast of the original SPC of Fig. 9.1A, the power flow direction is from right to left. The original source V_{s1} is then replaced by the load R2. Looking at the circuit in Fig. 9.1C, we see that it is nothing more than a mirror image of a buck converter topology! The process steps illustrated in Fig. 9.1 are a special case of duality and are referred to as *bilateral inversion*.

From this simple exercise, it becomes readily evident that boost and buck topologies are dual converter structures. It is interesting to observe that the boost converter circuit could have been easily "created" by applying the above processes to a buck converter circuit. For this reason, it is tempting to inquire which of these two SPCs is "more" fundamental than its counterpart. This particular question, when asked, often leads to much discussion with no definitive answer. For our purposes, we will assume that *both* are fundamental DC-DC power converters with duality relationships evident in their electrical characteristics.

The general rules for applying bilateral inversion can be stated as follows:

STEP 1. Each switch element is replaced by a diode phased to conduct current in the opposite direction from that seen by the switch element to be replaced.

STEP 2. Each commutating diode is replaced by a switch element phased to conduct current in the opposite direction from that seen by the diode to be replaced.

STEP 3. Conduction duty cycle of replacement switches is the inverse of that of original SPC switch elements (i.e., $D' = 1 - D$).

STEP 4. Assuming the input voltage to the original SPC circuit is V_s and its output voltage is V_o for a given value of D, the source voltage is replaced by a load with a potential V_s and the original load replaced by a source voltage with a potential V_o.

STEP 5. For the same value of D in Step 4, the replacement load and any output capacitance becomes:

$$R2 = R1 \left[\frac{V_o}{V_s} \right]^2, \quad C2 = C1 \left[\frac{V_s}{V_o} \right]^2.$$

STEP 6. For given values of V_s, V_o, D, and output power levels, the dual circuit will *not* change operating mode (i.e., continuous or discontinuous) over that of the original converter circuit.

9.2. INVERSION OF RECTIFIER NETWORKS

The principles behind bilateral inversion can be applied with equal facility to rectifier networks. For example, in Fig. 9.2, we have taken four common rectifier networks and applied the rules delineated earlier for bilateral inversion. As shown, the duals of these output rectifier circuits are the four common primary switch connections found in SPC's.

It is also interesting to note that the four rectifier connections shown in Fig. 9.2 represent only a few possibilities, with many other circuit arrangements in common use. This observation leads directly to the idea of applying bilateral inversion principles to all known rectifier connection arrangements to derive new and potentially useful switch connections.

Let us now explore this idea. The rectifier connection given in Fig. 9.2C is called a *voltage doubler*. Another circuit form of a single-winding voltage doubler exists and is shown in Fig. 9.3A. Applying the rules of bilateral inversion, the switch network of Fig. 9.3B results. This arrangement is an alternative to the common half-bridge connection and allows one side of the primary of T1 to be connected to the common return point of the source voltage. Although the need for capacitor C2 may not be apparent from Fig. 9.3A, practical SPC circuits seldom are designed to permit pulsating currents to flow in their power sources; therefore, C2 will normally be added to "smooth" out such current pulsations.

Voltage multipliers with gain ratios greater than 2 are possible and their networks can be inverted to evolve switch connections which exhibit the voltage and current characteristics of the multipliers. Fig. 9.4 shows a voltage quadrupler network (A) and its dual switch connection (B). In this switch network, the average voltage appearing across any open switch will be equal to 50% of the input source voltage (V_s)! As an example, the network of Fig. 9.4A would allow switch elements with breakdown ratings of 400 V to be used when operating from a 700 V source (some derating of breakdown voltage is assumed here). Therefore, series connection of switch elements to increase voltage breakdown capability is not required, as might be the case in conventional SPC primary switch connection schemes. Lin and Chua have demonstrated that this type of quadrupler (Ref. 3) has nine possible variations each of which will produce a switch connection with slightly different properties. A 7-times multiplier would have 115 variations, which should be enough to keep even the most ambitious analyst busy for some time.

There are other, and in some ways, better means for providing voltage multiplication by rectification and capacitive energy storage. One of the problems

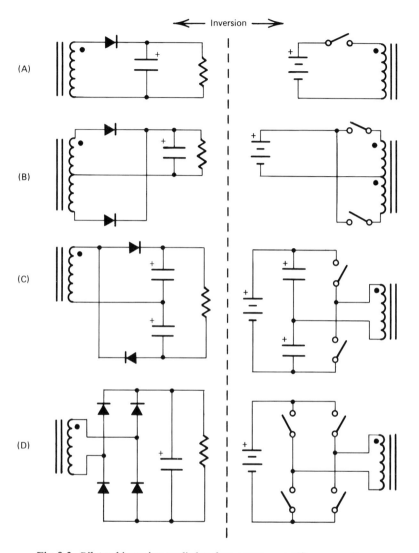

Fig. 9.2. Bilateral inversion applied to four common rectifier connections.

with the multiplier circuits just shown is the high output impedance and the large diode currents, both properties of which are transferred to their dual switch circuits. Other forms of voltage multipliers exist, some of which have much lower output impedances. The voltage doubler form of one of these rectifier connections is shown in Fig. 9.5A, with the dual switch connection in Fig. 9.5B. This latter circuit is a double-bridge switch connection in which each switch

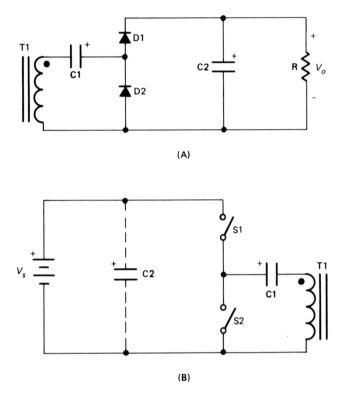

(A)

(B)

Fig. 9.3. Bilateral inversion of an alternate voltage-doubler rectifier connection.

sees a maximum OFF voltage of $V_s/2$ and a peak ON current equal to that of I_s. The transformer's peak-to-peak primary voltage will be equal to V_s. In this type of multiplier–derived switch connection, the switch currents do not increase in direct proportion to the multiplication number, as was the case for the connections in Figs. 9.2C, 9.3B, and 9.4B. The correct timing sequence for the SPC circuit of Fig. 9.5B is to close switch set S2, S3, S6, and S7 or switch set S1, S4, S5, and S8 simultaneously. In other words, each bridge set of Fig. 9.5B is operated with identical switching sequences.

Multipliers can be extended to even higher orders. Two examples of voltage triplers are illustrated in Fig. 9.6. In higher-order multiplier networks, there is a choice to be made in arrangement of energy storage capacitors. These capacitors may be connected in series (Fig. 9.6A), in which case each capacitor will see nearly the same value of DC bias. However, to keep the output impedance of the network low, the capacitors are usually large in value. In Fig. 9.6A, capacitors C1 and C2 should have values at least twice that of capacitors C3 and C4. Lower output impedance can be obtained if the energy storage capacitors

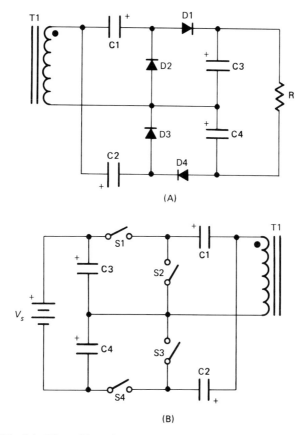

Fig. 9.4. Bilateral inversion of a quadrupler rectifier connection.

are connected in a parallel fashion as shown in Fig. 9.6B. Here, capacitors C3 and C4 will experience a DC bias of twice that appearing across C1 and C2. In a voltage-tripler network, high DC voltage across its capacitors is not usually a problem. However, in higher-order multipliers, capacitor sizes can become very large, since their stored energy is proportional to the square of the voltage appearing across their terminals. When multipliers of this variety are used for very high voltage gain, combinations of series and parallel capacitor methods are normally employed. Typically, these combinations consist of groups of parallel capacitor triplers or quadruplers stacked to form a series-parallel network system.

Another form of a low impedance voltage-doubler circuit is shown in Fig. 9.7A, where two separate transformer windings and associated rectifier bridge networks are tied in a series fashion. Since there are no series capacitance elements in this doubler, output impedance can be made very low at the expense of additional

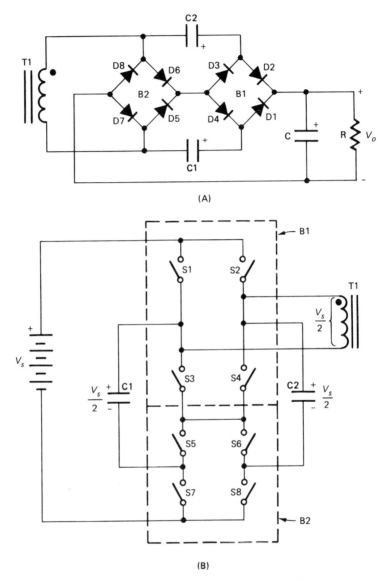

Fig. 9.5. Inversion of low impedance voltage multipliers (I).

transformer complexity. The output voltage of this system can be made even higher in value by adding more transformer windings and bridge networks without the loss of the low output impedance characteristic. For this reason, this particular voltage multiplier approach is very frequently used by designers in converters that must have high voltage output capability.

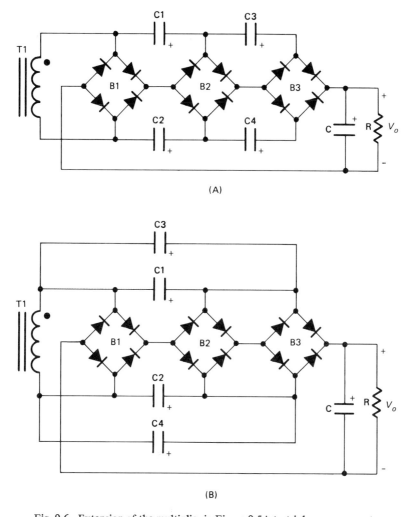

Fig. 9.6. Extension of the multiplier in Figure 9.5A to tripler arrangements.

The corresponding dual switch connection of this multiplier is illustrated in Fig. 9.7B. This is another double-bridge scheme (B1, B2) in which the maximum OFF voltage of each switch is $V_s/2$, with peak switch ON currents equal to that of the average source current. This switch connection would be particularly useful in an SPC which must operate from a 1000 to 1200 V voltage source and process tens of kilowatts of output power. Semiconductor switch elements exist today that could be used in the circuit of Fig. 9.7B to easily provide 50 kW or more output power. Even higher source voltage values and output power levels could be accommodated by stacking additional switch bridges and transformer

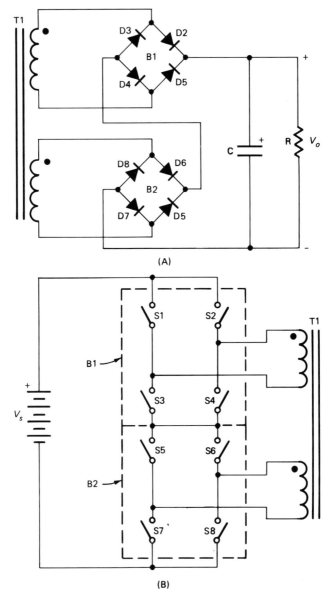

Fig. 9.7. Inversion of low impedance voltage multipliers (II).

windings to the extent that an SPC could conceivably be designed to provide hundreds of kilowatts of output power.

In addition to the vast number of voltage-multiplier connection possibilities, there are an equally large number of networks associated with polyphase rectifica-

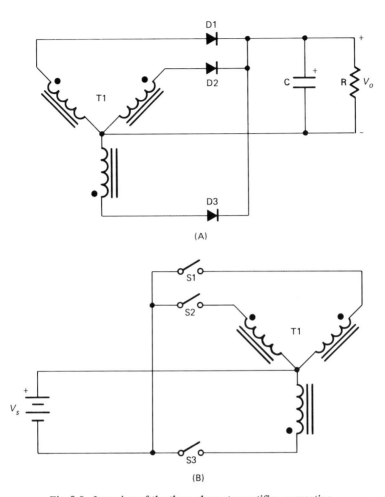

Fig. 9.8. Inversion of the three-phase star rectifier connection.

tion to which the principles of bilateral inversion can be applied. Two circuit examples and their duals are shown in Fig. 9.8 and 9.9. Polyphase rectifier networks provide many electrical benefits, especially at high power-processing levels. These benefits include higher ripple frequency, automatic current-sharing among multiple devices, better copper utilization of associated transformer windings, and many more. To a great extent, these desirable electrical properties are retained in the dual polyphase switch connections. For example, if the primary switches shown in Fig. 9.9B are operated at a switching rate of 20 kHz, the output voltage ripple frequency of 'any transformer secondary winding will be (6) (20 kHz), or 120 kHz. Obviously, any low-pass output filter will be smaller in size because of the higher ripple frequency. The timing sequence of switching

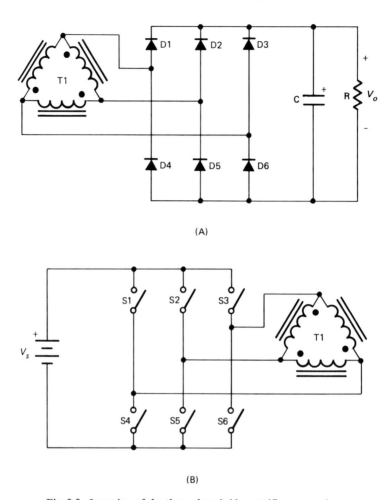

(A)

(B)

Fig. 9.9. Inversion of the three-phase bridge rectifier connection.

for this polyphase converter is diagrammed in Fig. 9.10. We observe that maximum duty cycle of any switch is approximately 33%, whereas, in a single-phase full-bridge connection, the duty cycle limit would be 50% Also evident is the fact that *only* one switch at a time will turn on if $D < \frac{1}{6}$. This would imply that the output voltage from the converter must be zero; thus, the required range of the duty cycle of any switch is $\frac{1}{6}$ to $\frac{1}{3}$ in order to produce zero-to-full output voltage. In the polyphase full-bridge connection, inherent delays in switching actions (such as storage time in BJTs) do not pose conduction overlap problems under converter overload conditions (i.e., small values of D). As we have just seen, such delays can be as long as one-sixth of the entire switching interval (T_s) and still be tolerated.

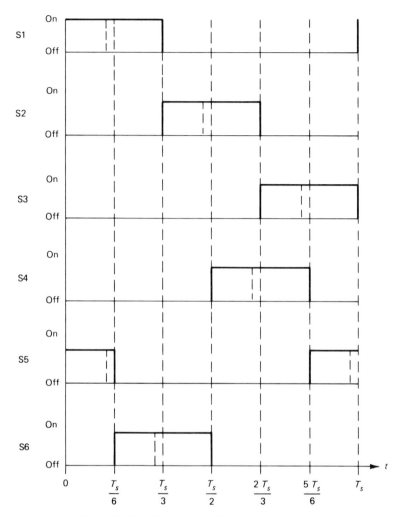

Fig. 9.10. The three-phase bridge switching sequence.

By equipping the polyphase switch connection of Fig. 9.9B with the multiple bridge approach of Fig. 9.7, very high power converters are possible and practical.

9.3. TOPOLOGY MODIFICATION BY SWITCHING SEQUENCE VARIATIONS

Up to this point, we have assumed that each SPC topology has individual and distinguishable sets of electrical characteristics, i.e., buck, boost, or some combination thereof. Now we will look at SPCs whose topologies are still fixed in circuit form, but whose conversion characteristics can be drastically altered by changing the sequence in which their switches are opened or closed. As an example of such systems, the circuit of Fig. 5.8 will be used. In Fig. 9.11, we see the two "normal" switching states of this converter. At the beginning of the switching cycle, switches S1 and S2 are closed (State I) and the current flows from the source through inductor L1 and switch S1 to the ouput via the transformer, T1. Part way through the first half-cycle of switching operation, S1 is opened (State II) and energy stored in L discharges through D1 and S2 into the output. This two-state sequence is then repeated using S3 and S4 to complete one switching cycle of operation. As was shown in Chapter 5, the SPC of Fig. 9.11 is a buck-derived converter.

An alternative two-state switching sequence is shown in Fig. 9.12. At the beginning of a switching cycle, S1, S2, and S3 are closed (State I). Once again, current flows from the input source through L. However, now the average inductor current divides equally between S1 and S3 and effectively places a "short circuit" across the primary of T1. Therefore, no energy is delivered to the out-

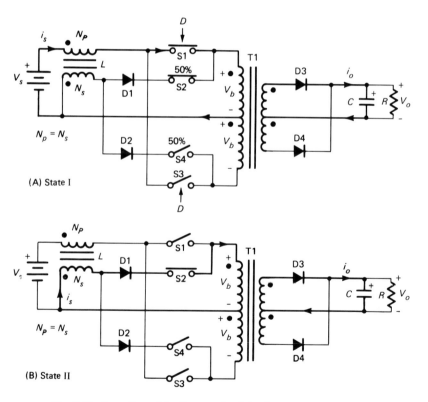

Fig. 9.11. Operation of the Severns circuit in the buck-derived mode.

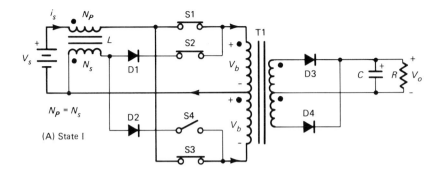

(A) State I

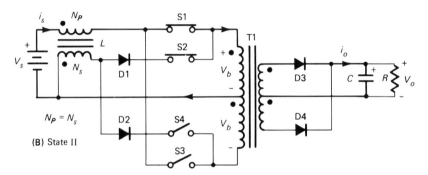

(B) State II

Fig. 9.12. Operation of the Severns circuit in the boost-derived mode.

put during State I. Part way through this half-cycle of conversion action, S3 opens (State II) and all of the energy in L is delivered to the output through S1. Notice that the secondary winding of inductor L, as well as S2, S3, D1, and D2 play *no* part in conversion operation using this timing sequence. Therefore, this SPC operates in much the same manner as did the boost-derived converter shown in Fig. 6.4B.

The difference between the converter switching sequences of Fig. 9.11 and Fig. 9.12 lies in switch duty cycle D. In Fig. 9.11, D is always less than 50% while in Fig. 9.12, it is always greater than 50% Thus by simply altering switch duty cycle, this SPC can operate either as a buck or a boost converter! Another circuit variation that would provide a similar type of operation with proper sequencing of switches is the full-bridge SPC of Fig. 5.17A.

This is a rather peculiar and interesting capability and one may wonder if sequence changes in SPC timing have any practical application. To demonstrate the utility of this technique, an illustration is in order. Some time ago, one of the authors was asked to design a high power off-line (i.e., source voltage is derived from high-voltage rectified AC) unity-power-factor converter with a very high quality input current waveform. The unity power factor requirement clearly indicated that a boost-derived SPC was needed, since all buck-derived converters have pulsating input currents. The SPC circuit topology chosen was the full-bridge overlapping-conduction converter of Fig. 6.10A. The converter worked very well for normal steady-state operation. However, because it was a boost-derived SPC, it could not, by its very nature, provide inrush and overload current-limiting, as was pointed out in Chapter 3. To circumvent these turnon and overload problems, the SPC topology circuit was modified to that shown in Fig. 9.13. Here, an auxiliary winding was added to the input inductor L and connected to the center tap of the primary on T1 through a diode (D1). This modification allowed the converter to operate in a "buck" mode when overload or turnon inrush current control protection was required. The circuit is affectionately called the *shower converter* because the changes in topology were evolved early one morning while one of the authors was standing under a hot shower, with its warm water apparently acting as a catalyst in the innovation process.

The moral of this story brings us to another key idea relative to SPC design.

KEY IDEA #6

IF YOU WANT THE PROPERTIES OF A BUCK CONVERTER, THEN THE SPC *MUST* BE BUCK-DERIVED. IF THE PROPERTIES OF A BOOST CONVERTER ARE REQUIRED, THEN THE CIRCUIT CHOSEN *MUST* BE BOOST-DERIVED. IF *BOTH* BUCK AND BOOST PROPERTIES ARE DESIRED OF AN SPC, THEN *EITHER* THE CIRCUIT MUST BE EVOLVED FROM A COMBINATION OF BUCK

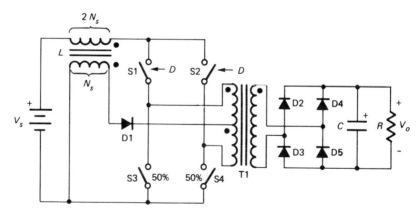

Fig. 9.13. The Shower circuit.

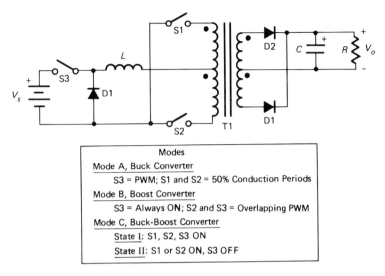

Modes
Mode A, Buck Converter
    S3 = PWM; S1 and S2 = 50% Conduction Periods
Mode B, Boost Converter
    S3 = Always ON; S2 and S3 = Overlapping PWM
Mode C, Buck-Boost Converter
    State I: S1, S2, S3 ON
    State II: S1 or S2 ON, S3 OFF

Fig. 9.14. The TRW converter operating in three different switching modes.

AND BOOST-DERIVED CONVERTERS *OR* THE SPC *MUST* BE ABLE TO ALTERNATE BETWEEN BUCK AND BOOST-DERIVED MODES OF OPERATION.

We can carry the process of altering the timing sequence even further to provide three modes of conversion operation as shown in Fig. 9.14. Here, by varying the switching sequence, a converter, shown earlier in Chapter 5 (Fig. 5.3A) can assume the characteristics of a buck, a boost or a buck-boost SPC.

9.4 APPLYING BILATERAL INVERSION TO COMPLEX CONVERTERS

The rules for bilateral inversion can be applied directly to more complex SPCs to evolve their duals and, in some cases, new converter structures will result. For example, Fig. 9.15A shows the parallel quasi-squarewave converter circuit of Chapter 4. Applying the rules of bilateral inversion to this SPC, the topology shown in Fig. 9.15B evolves and, by inspection, is the parallel overlapping-conduction boost converter of Chapter 5. Again, the existence of duality between contemporary buck and boost-derived converters becomes apparent.

In Fig. 9.16, we see that the dual of another buck-derived converter of Chapter 5 (Fig. 5.9A) is a flyback SPC in series with the output of a DC transformer (Fig. 9.16B). The input current of this dual topology is continuous in form, while its output current is discontinuous. These properties are to be expected since the dual of the SPC in Fig. 9.16A must be a boost-derived converter.

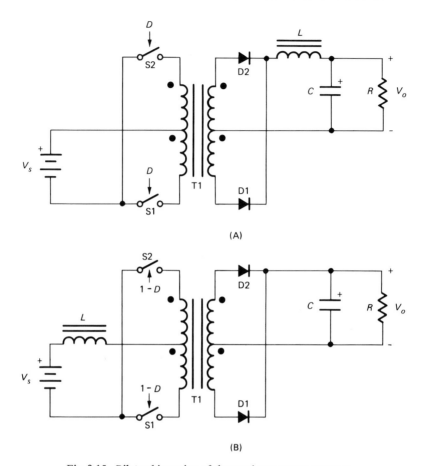

Fig. 9.15. Bilateral inversion of the quasi-squarewave converter.

Finally, in Fig. 9.17, we see that the dual to a boost-derived converter with a tapped primary winding (Fig. 8.20B) is a tapped secondary winding buck-derived SPC system as shown in Fig. 9.17B.

At this point, it is evident that, *for every known buck-derived converter, there exists a dual boost-derived converter, and vice versa.* Therefore, should new SPC topologies be conceived, their dual structures can easily be evolved using the methods of bilateral inversion. However, in some cases, a bilaterally-inverted network is identical to the original parent circuit. The Ćuk and buck-boost converters are prime examples of such exceptions. While these two particular converters have identically dual SPCs that result from bilateral inversion manipulation, we will soon see that applying general duality principles to these two SPCs will finally reveal the dual nature of their topologies.

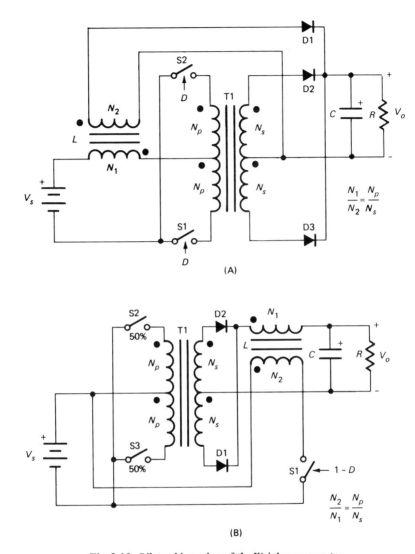

Fig. 9.16. Bilateral inversion of the Weinberg converter.

The previous three examples of this discussion concerned themselves with single-output converter structures. When bilateral inversion principles are applied to single-input multiple-output converter systems, the resultant dual SPCs will have a single-output and a multiple-input capability.

While converters with the capability of accepting inputs from more than one source of power are occasionally useful, the general rule for effective application of the principle of bilateral inversion is to reduce the SPC circuit to a single in-

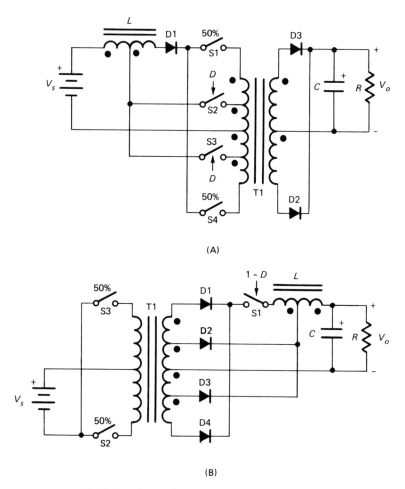

(A)

(B)

Fig. 9.17. Bilateral inversion of the Hughes circuit.

put and a single output, perform the inversion processes, and then arrange the number of inputs and outputs to suit the application need.

9.5 APPLYING THE GENERAL DUALITY PRINCIPLE TO SPC CIRCUITS

In the first four sections of this chapter, we have been discussing a very special and limited case of duality namely, bilateral inversion. In this section, we will expand our discussions to include the general concept of duality and how to apply it to SPC circuits.

The duality principle is a fundamental concept in our understanding of nature and its application to electrical circuits is well established. However, most discussions of duality are limited to linear circuits or those with rather simple nonlinearities. Until the work by Ćuk [5], it was not clear how, or even if, it was possible to apply duality to SPC circuits, which are highly nonlinear. As we shall see, duality in its general sense can be applied to this class of circuits. Applying duality will provide us with an analytical method of simplifying the derivation of the properties of many converters. The insight gained into the relationships between SPC circuits will enable us to better judge the applicability of different converters.

Basically, Dr. Ćuk's approach is to reduce a converter circuit to a series of linear equivalent circuits for each switching state, much as we did when deriving the averaged small-signal models of Chapter 2. A dual network for each circuit state is then developed, and the resulting networks recombined into a single circuit. For many of us (the authors included), the principles behind duality of electrical networks were subjects introduced briefly in a college sophomore or junior network analysis course with little or no application in subsequent day-to-day work. So, before charging into an exposition of using duality in SPC circuits, we will stop to review the duality principle and related procedures for generating circuit duals. Of necessity, the subject discussions to follow are limited in scope, but most basic electrical network analysis textbooks provide more rigorous treatments for those readers who wish to delve deeper into duality principles.

In electrical networks, duality is usually limited to those linear circuits whose graphs are planar. A *graph G* is said to be *planar* if it can be drawn on a plane in such a way that no two branches cross at a point which is *not* a node. For example, the graph in Fig. 9.18A is planar while that of Fig. 9.18B is not.

In a planar graph, any closed loop which has no branches in its interior is called a *mesh*. In Fig. 9.18A, loops *a-b-e-a*, *c-d-e-c*, and *b-c-e-b* are meshes. A loop which contains no branches in its exterior is called an *outer mesh*. With these basic definitions in mind, we can now move to define what is meant by duality in graphs and then extend it to electrical circuits.

A graph *G** is said to be the dual of another graph *G* if the following three conditions are satisfied:

CONDITION 1. There is a one-to-one correspondence between all meshes of *G* and the nodes of *G**.

CONDITION 2. There is a one-to-one correspondence between all nodes of *G* and the meshes of *G**.

CONDITION 3. There is a one-to-one correspondence between the branches of each graph such that, whenever two meshes of one graph have

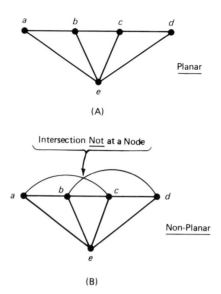

Fig. 9.18. Examples of planar and nonplanar graphs.

a branch in common, the corresponding nodes of the other graph have the corresponding branch connecting these nodes.

In analyzing electrical circuits, the direction of the assumed current flow in each branch is usually indicated. The graph of such a circuit is called an *oriented graph*, with a current direction indicated on each of its branches. For an oriented graph G, where each branch current has a reference direction, the orientation of the dual graph G^* can be obtained by rotating the current reference direction of each branch by $90°$ *clockwise* to obtain the current direction of the corresponding branch in G^*.

All of these rules sound very complex and formal, but as we see in Fig. 9.19, the actual process is quite simple. Here, an oriented graph is shown in Fig. 9.19 A. In Fig. 9.19B, we place a node inside each mesh including the outer mesh (nodes 1^*, 2^*, 3^*, and 4^*). Branches are then connected between the new nodes so that one branch of the new graph intersects each of the branches of the original one. We now have the dual graph as shown in Fig. 9.19C. From each intersection in the dual graph the orientation of the corresponding branch is then determined. For analysis convenience, it is common practice to draw similar circuits with the same layout format. In doing so, it is much easier for a designer to recognize the type of circuit being examined. It is quite possible to draw common converter circuits in a manner where their structures become unrecognizable. Sometimes, the process of generating the dual of a network results in a circuit

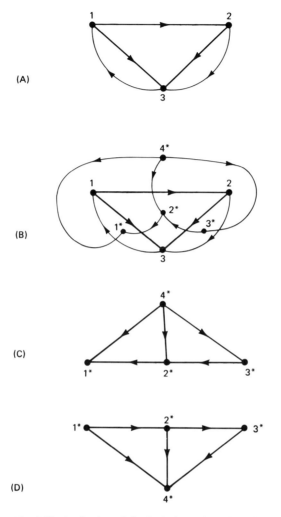

(A)

(B)

(C)

(D)

Fig. 9.19. Derivation of the dual of an oriented graph.

form which is not "standard" or easily recognizable. Usually simple drawing rotation or translation techniques will place the circuit in the proper perspective; however, it is often necessary to also mirror-image the circuit. An example of mirror-imaging is shown in Fig. 9.19D. Notice that the reference current direction of each branch of the mirror image must be reversed.

In an electrical network, each of its branches will contain circuit components with particular electrical characteristic, i.e., a resistor, an inductor, a current source, etc. Duality requires that the following transformations be made:

$$v \rightarrow i^*, \quad i \rightarrow v^*, \quad q \rightarrow \phi^*, \quad \phi \rightarrow q^* \tag{9.1}$$

where v = branch voltage, i = branch current, q = branch charge, and ϕ = branch flux variation.

From elementary circuit theory, we know that for each branch:

$$\left\{ \begin{array}{ll} v = iR & v^* = i^*R^* \\[2ex] v = L\dfrac{di}{dt} & v^* = L^*\dfrac{di^*}{dt} \\[2ex] i = C\dfrac{dv}{dt} & i^* = C^*\dfrac{dv^*}{dt} \end{array} \right\} \tag{9.2}$$

From the relationships of Eq. (9.2), it can be shown that the corresponding branch duality relationships in Table 9.1 are true. The dual relationship for D^* is demonstrated in the next section of this chapter.

Let us now apply the relationships of Table 9.1 and the dual graphing procedure to the network N in Fig. 9.20A. The resulting dual network N^* (mirror-imaged) is shown in Fig. 9.20C.

The duality principle is actually much more general than the simple procedures outlined here would indicate. The principle of duality applied to dual networks can be stated as follows:

Consider an arbitrary planar network N and its dual N*. Let S be any true statement concerning the behavior of N. Let S* be the statement obtained from

**TABLE 9.1. Branch Component
Duality Relationships.**

Quantity	Dual Element
v	i^*
i	v^*
q	ϕ^*
ϕ	q^*
R	$R^* = G = 1/R$
G	$G^* = R = 1/G$
C	L^*
L	C^*
Open circuit	Short circuit
Short circuit	Open circuit
D	$D^* = 1 - D$
Voltage generator	Current generator
Current generator	Voltage generator

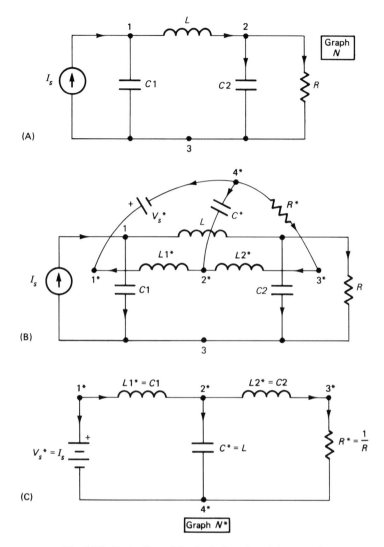

Fig. 9.20. Derivation of the dual of an electrical network.

S by replacing every graph-related word or phrase (node, loop, mesh, etc.) by its dual and every electrical quantity (voltage, current, impedance, etc.) by its dual. Then S* is a true statement concerning the behavior of N*.

Before proceeding to applying duality to SPC circuits, we need to take another look at the elementary buck and boost converter circuits. In Chapters 2 and 3, all of the basic converter circuits were shown with a voltage source at their in-

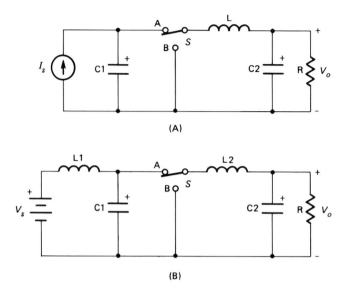

Fig. 9.21. The buck converter fed from a current source.

puts. We could also have used a current source as shown in Fig. 9.21A. Here, capacitor C1 has been added across the input current source so that the input voltage does not rise to infinity when switch S is in the B position. As a practical matter, SPCs very often have a "current source" at their inputs because of the common requirement for limiting the amount of conducted EMI noise seen by their input voltage sources. A typical example is shown in Fig. 9.21B where, to obtain a high degree of isolation from input current pulsations, an input low-pass filter (L1, C1) is added. The inductance of L1 is usually large and its current is essentially DC, since C1 provides the dynamic energy source for the SPC. This arrangement acts very much like a voltage-limited current-source. In Fig. 9.22, the four combinational possibilities of source and elementary converter topologies are indicated. Notice that we have tinkered a bit with the SPC circuits. In the buck converters (Fig. 9.22A and B), the output capacitors have been eliminated since their presence is not required for proper converter function, as was pointed out in Chapters 1 and 2. However, in the boost converters (Fig. 9.22C and D), the output capacitor (C2) is required and so it has been retained. Another change indicated is the addition of a capacitor (C1) across each of the voltage sources. From a circuit function point of view, this capacitor serves no real purpose, since the voltage sources are assumed to have zero impedance. The reason for adding C1 is a matter of practicality in the application of duality principles to SPC systems. As we will soon see, its presence will lend credibility to the existence of many dual SPCs. The reader may notice that, by removing

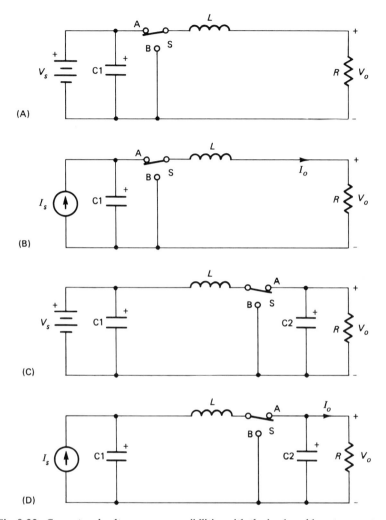

Fig. 9.22. Current and voltage source possibilities with the buck and boost converters.

the loads and sources from the circuits of Fig. 9.21, the buck circuits are mirror images of the boost circuits. In order to apply duality to SPC circuits and obtain meaningful results, it is necessary to manipulate the circuits in this manner.

Looking again at Fig. 9.22, we see that circuit A is a normal buck converter since $V_o/V_s < 1$, this SPC can be called a *buck-voltage* converter. For circuit B, $I_o/I_s > 1$, hence, we can call this SPC a *boost-current* converter. Circuit C is a *boost-voltage* converter, since $V_o/V_s > 1$, but circuit D is a *buck-current* converter since $I_o/I_s < 1$. We will use these four names in the duality discussions to follow, but we will not expand the number of basic converter types from two to four since there seems to be no practical or advantageous reason to do so, at this time at least. Nearly all converters in use today are voltage converters that provide regulated output voltages. Applications where the primary function of the converter is to provide a regulated output current do exist, but are relatively low in number. For such rare applications, it would be interesting and perhaps very useful to restate the voltage-converter-oriented presentations of this book in terms related to current conversion. The duality principle is the most convenient and efficient means to do this. However, because of our stated limitation of scope of subject matter, the authors leave current-conversion examinations as duality exercises for the reader.

We are now ready to apply duality methods to some of the common converter circuits of earlier chapters.

9.6 APPLYING DUALITY TO SPC CIRCUITS

Consider the circuit in Fig. 9.23A which is a current-boost converter or, if you prefer, a buck-voltage converter with a voltage-limited current input source. If we presume the continuous mode of operation, then there are two equivalent linear circuits that correspond with the two switch positions A and B. The circuits resulting from these two switch states are shown in Fig. 9.23B. Since these circuits are linear, we can apply the principles of duality directly using the rules and procedures developed in the previous section of this chapter. This is indicated in Fig. 9.23B, with the final results shown in Fig. 9.23C. The two dual linear circuits can now be reunited with switch S positioned as shown in Fig. 9.23D. Note that this SPC is a boost converter. Once again, it becomes apparent that the boost converter is the dual of the buck converter, just as we found when bilateral inversion techniques were applied. By convention, switch conduction duty cycle (D) in the buck converter is defined as the portion of the total switching interval (T_s) that switch S is in position A. However, for the boost converter, D is the portion of T_s that S is in the B position. From this observation, we can immediately see the origin of the dual transformation definition, $D^* = 1 - D$, given earlier in Table 9.1. These duty cycle conventions are simply the result of the normal realization of the SPDT switch by the use of a diode and a transistor and subsequently defining D as the ON time of the transistor.

For the current source buck regulator of Fig. 9.23A, it can be shown that:

$$\frac{I_o}{I_s} = \frac{1}{D} . \tag{9.3}$$

From the rules for duality developed earlier, we know that:

$$I_o \rightarrow V_o^* \tag{9.4}$$

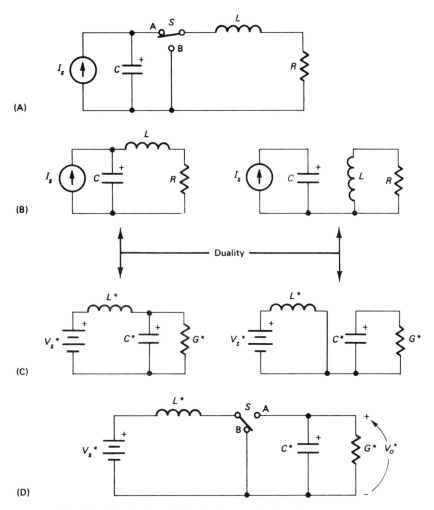

Fig. 9.23; The duality relationship between buck and boost converters.

$$D \rightarrow D^* = 1 - D \qquad (9.5)$$

$$I_s \rightarrow V_s^*. \qquad (9.6)$$

Therefore:

$$M = \frac{V_o^*}{V_s^*} = \frac{1}{1 - D}. \qquad (9.7)$$

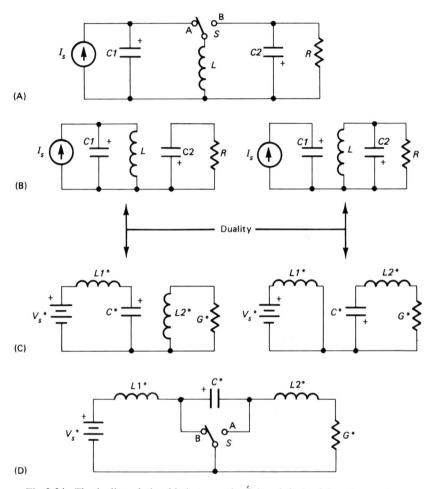

Fig. 9.24. The duality relationship between the Ćuk and the buck-boost converters.

We have already established the gain relationship indicated by Eq. (9.7) in Chapter 3 by a completely separate (and more involved) circuit analysis method. We could have applied the principle of duality to the buck converter of Chapter 2 and derived the characteristics of the boost converter directly with much less analysis effort. This example illustrates the power of the duality principle.

We can now introduce the following key idea related to SPC characteristics and duality:

KEY IDEA #7

IF WE KNOW THE PROPERTIES OF A GIVEN SPC CIRCUIT, WE CAN DERIVE THE PROPERTIES OF THE DUAL OF THE

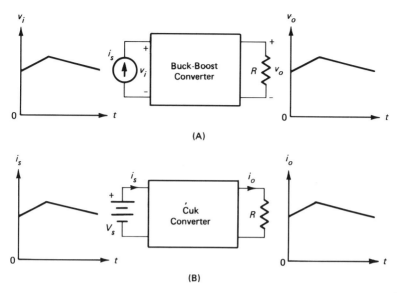

Fig. 9.25. Duality of the input and output voltages and currents in buck-boost and Ćuk converters.

SPC BY APPLYING THE RULES OF DUALITY TO THE ORIGINAL CONVERTER PROPERTIES. ALL OF THE PROPERTIES OF THE DUAL CONVERTER WILL BE DUALS OF THOSE OF THE ORIGINAL SPC.

Let us now apply duality to a more complex converter. Fig. 9.24A shows a buck-boost converter with a current source input. By following the duality steps indicated in Fig. 9.24B and C, we arrive at the dual converter of Fig. 9.24D, which is none other than the Ćuk converter. The Ćuk and the buck-boost are therefore dual SPCs. Interestingly enough, if the Ćuk converter approach had not been conceived by quite a different process (see Chapter 7, Fig. 7.6) it would have appeared quite naturally from the application of duality principles to the buck-boost converter.

In Fig. 9.25, note that the nonpulsating nature of input and output voltage waveforms in the buck-boost converter is transferred to the input and output currents in the dual Ćuk converter. The nonpulsating nature of the input and output currents is a major advantage of the Ćuk converter.

9.7 DUALITY IN INDUCTIVE AND CAPACITIVE ENERGY TRANSFER

In earlier chapters, we made much ado about the central role played by inductor current in determining the operating mode of an SPC. In most converters, it is the inductor which transfers the energy from the input to the output. The Cuk converter is unique in this respect in that it uses a capacitor to transfer energy from the input to the output as well as inductive energy storage. From the principle of duality, we know that the dual to an inductor current will be a capacitor voltage. In the Cuk converter, the capacitor voltage can be either continuous or discontinuous. Since the capacitor is an active part of this SPC's energy transfer mechanism, the conduction mode of the capacitor voltage is just as fundamental in determining the converter properties as its inductor's conduction modes. Readers interested in a more detailed discussion of the phenomena of discontinuous capacitor voltage in a Ćuk converter are referred to Ref. 5 in the bibliography of this text.

10. Small-Signal Models of Complex Converters and Regulators

In Chapters 2 and 3, a great deal of effort was expended in the generation of equivalent circuit as well as mathematical models for the basic buck and boost converters. In subsequent chapters, a multitude of SPC circuits were derived, for the most part, from these two elementary converters. Throughout the development of these more complex SPCs, we have been very careful to keep track of their basic constituents, i.e., whether they were buck-derived, boost-derived, or some combination thereof. One of the many reasons for carefully identifying their basic constituents is that by doing so, we can quickly determine which of the basic SPC models of Chapters 2 and 3 apply to any given converter. As will be evident in this chapter, this recognition capability greatly reduces the effort in analyzing a given SPC circuit.

Up to this point, we have treated all of the SPC circuits by themselves and ignored the fact that most converters are normally used in conjunction with other electronic networks to realize a voltage-regulating system. An important exception to this viewpoint was the development of small-signal circuit models that provided for external dynamic control of duty cycle. In this chapter, we will present some SPC examples in which voltage regulation and EMI filter networks are included. These examples will serve to demonstrate the flexibility and the power of small-signal modeling techniques for SPC systems.

10.1. THE CANONICAL MODEL

In Chapters 2 and 3, small-signal models for the buck and boost converters were developed for two distinct conduction modes. The conduction modes were defined as continuous and discontinuous, and reflected the nature of the current waveform within the inductor of these SPCs. For each conduction mode, we found that the small-signal describing equations and the equivalent circuit models had the same form for both buck and boost converters. The differences between each SPC model were the component values of network elements of the circuit model and corresponding constants in the circuit equations. These observations bring us to a key generalization for modeling SPCs:

Key Idea #8

THE SMALL-SIGNAL CIRCUIT MODEL OF ANY SPC FOR A *GIVEN CONDUCTION MODE* CAN HAVE THE SAME TOPOLOGY AND TYPE OF CIRCUIT ELEMENTS. THE VALUES ASSIGNED TO EACH CIRCUIT ELEMENT OF THE COMMON MODEL TOPOLOGY ARE DEPENDENT ON INDIVIDUAL CONVERTER PROPERTIES.

This generalization allows us to make a rather spectacular simplification in the analysis of SPC circuits, since we will now know beforehand what the structure of the model will look like. The analysis problem then reduces to one of calculation of model element values using standard techniques of linear circuit theory.

Our general small-signal model will be called the *canonical model* and is patterned after that conceived by Drs. Middlebrook and Ćuk (Ref. 6). By necessity, separate canonical models for each conduction mode are required. The canonical model for the continuous mode is given in Fig. 10.1. For convenience, various element impedances or characteristics are expressed in terms of complex frequency, i.e., the s domain. To further simplify the equations of this presentation, we will omit the explicit expression of s in such instances and let it be implicit. For example, $d(s)$ will be written as d.

We see that the model of Fig. 10.1 is divided into three sections, each representing the functions inherent in an SPC. The first box represents the small-signal control properties due to $\hat{d}$, the second box represents the inherent DC

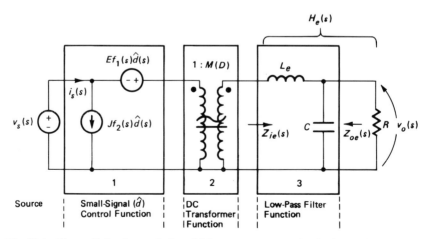

Fig. 10.1. The small-signal canonical model for converter operation in the continuous mode.

transformation function of the converter to be modeled $(M(D))$, and the third box represents the effective output low-pass filter network of the SPC. As defined in Chapters 2 and 3, M is the ideal ratio of V_o/V_s for the SPC to be modeled. The correction factor that accounts for the *nonideal* value of M (i.e., efficiency of power conversion), is included in the transfer function, H_e. The filter network, shown in the third box of Fig. 10.1, represents only one of an infinite variety of possibilities. The only constraints placed on this filter network are the need to exhibit a *low-pass* characteristic and, of course, that it correctly represent the effective filter circuit elements of the converter under consideration. As we shall see in the next section of this chapter, this equivalent low-pass filter network can become quite complex. Like any linear filter circuit, it can be characterized by its input-to-output voltage-transfer ratio (H_e), its input impedance (Z_{ie}) and its output impedance (Z_{oe}). Note that the converter's load (R) must be included in the determination of H_e and Z_{ie}. Also, any source impedances must also be included in Z_{oe}, suitably transformed through the DC transformation function of the second box in Fig. 10.1. Having a single model such as this for *all* converters can be extremely useful in the day-to-day design of SPC circuits. The following analysis advantages come immediately to mind.

First of all, an ever-increasing number of designers are using computer-aided methods to perform circuit analyses. By using a single universal model topology like the canonical form of Fig. 10.1, it is only necessary to write one computer analysis program for the model topology, inserting different values for the circuit elements of the model to examine the small-signal characteristics of *any* converter desired.

Second, the differences in small-signal characteristics of converters can easily be examined by comparing associated canonical model element values. For example, the range of D can be derived directly from $MH_e(0)$ (i.e., the product of M and the DC transfer function of the output filter), the range of input voltage (V_s), and variation in load resistance (R). Another example would be the comparison of control-to-output transfer functions of buck and boost SPCs. Here, the value of $f_1(s)$ in Fig. 10.1 will be equal to 1 for the buck converter. However, in the boost model, $f_1(s)$ will be complex and contain a *right* half-plane zero.

Thirdly, low-pass filter networks of SPC models can be compared in order to estimate their relative filtering capability for a given amount of signal attenuation desired.

There remains, of course, the analysis problem of determining canonical model element values for an SPC. There are two ways one can approach this problem. One method is to manipulate the equivalent circuit model and, very often, this can be done by simple inspection. In the next section of this chapter, we will give several examples of this inspection procedure. The second approach for determining model element values is to directly manipulate the circuit equations that describe features of an SPC. Looking at the canonical model in Fig. 10.1, we can write by inspection that:

$$\hat{v}_o = (\hat{v}_s + Ef_1 \hat{d}) MH_e \tag{10.1}$$

$$\hat{i}_s = Jf_2 \hat{d} + (Ef_1 \hat{d} + \hat{v}_s) \frac{M^2}{Z_{ie}} . \tag{10.2}$$

Eqs. (10.1) and (10.2) can be rearranged so that:

$$\hat{v}_o = (MH_e) \hat{v}_s + (Ef_1 MH_e) \hat{d} \tag{10.3}$$

$$\hat{i}_s = \left(\frac{M^2}{Z_{ie}} \right) \hat{v}_s + \left(Jf_2 + \frac{M^2 Ef_1}{Z_{ie}} \right) \hat{d} . \tag{10.4}$$

These equations can be further simplified by stating that:

$$\hat{v}_o = G_{vs} \hat{v}_s + G_{vd} \hat{d} \tag{10.5}$$

$$\hat{i}_s = G_{is} \hat{v}_s + G_{id} \hat{d} \tag{10.6}$$

Using Eqs. (10.3) through (10.6), we can derive expressions for the canonical model circuit elements:

$$Ef_1 = \frac{G_{vd}}{G_{vs}} \tag{10.7}$$

$$Jf_2 = G_{id} - Ef_1 G_{is} \tag{10.8}$$

$$H_e = \frac{G_{vs}}{M} . \tag{10.9}$$

The ideal SPC voltage gain function has already been determined as:

$$M = \frac{V_o}{V_s} \ \ \text{(IDEAL)} . \tag{10.10}$$

Further, if we define f_1 and f_2 such that:

$$f_1(0) = f_2(0) = 1, \tag{10.11}$$

then expressions for both E and J in the model can be readily determined.

The procedure for determining the canonical model element values is quite straightforward and can be stated as follows:

STEP 1. Derive the SPC equations using state-space averaging procedures, being careful to include $\hat{v}_o$ and $\hat{i}_s$ in any resulting expressions.

STEP 2. Rearrange the circuit equations of Step 1 so that they assume the form of those shown by Eqs. (10.5) and (10.6).

STEP 3. Solve for the element values of the small-signal model using Eqs. (10.7) through (10.11).

By way of an example of this three-step procedure, we will now develop the small-signal canonical model for the ideal buck-boost converter shown in Fig. 10.2A. Notice here the output voltage (v) has been defined as "positive" with respect to the common return of the SPC, even though the DC output voltage of the buck-boost converter is actually inverted (see Chapter 7). The "positive" convention is used here to be consistent with the format of the canonical model and, if our procedure is valid, the SPC's inversion property should appear in the final model as a natural consequence. The continuous mode will be assumed so that the converter must have two switching states, as shown in Figs. 10.2B and C. Applying the state-space averaging procedure to these two networks, we find from the DC relationships that:

$$M = -\frac{D}{D'} \tag{10.12}$$

which clearly indicates the inverting nature of the converter.

The small-signal equations of the averaged SPC are:

$$\frac{d\hat{i}}{dt} = \left[\frac{D'}{L}\right]\hat{v} + \left[\frac{D}{L}\right]\hat{v}_s - \left[\frac{V}{LD}\right]\hat{d} \tag{10.13}$$

$$\frac{d\hat{v}}{dt} = -\left[\frac{D'}{C}\right]\hat{i} - \left[\frac{1}{RC}\right]\hat{v} - \left[\frac{V}{D'RC}\right]\hat{d} \tag{10.14}$$

$$\hat{v}_o = \hat{v} \tag{10.15}$$

$$\hat{i}_s = \hat{i}D - \left[\frac{V}{RD'}\right]\hat{d}. \tag{10.16}$$

We can now transform these equations into the complex frequency domain and rearrange them into the forms of Eqs. (10.5) and (10.6). As a result of this transformation:

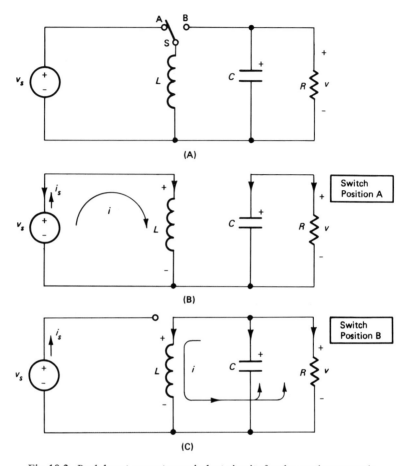

Fig. 10.2. Buck-boost converter equivalent circuits for the continuous mode.

$$\hat{v}_o = \left(-\frac{D}{D'}\right)\left[\frac{1}{1 + s(L_e/R) + s^2 L_e C}\right]\hat{v}_s$$

$$+\left(\frac{V}{DD'}\right)\left[\frac{1 - s(L_e D/R)}{1 + s(L_e/R) + s^2 L_e C}\right]\hat{d} \qquad (10.17)$$

$$\hat{i}_s = \left[\frac{D^2}{R(D')^2}\right]\left[\frac{1 + sRC}{1 + s(L_e/R) + s^2 L_e C}\right]\hat{v}_s$$

$$-\left[\frac{V}{R(D')^2}\right]\left[\frac{1 + D + sRC}{1 + s(L_e/R) + s^2 L_e C}\right]\hat{d} \qquad (10.18)$$

where

$$L_e = \frac{L}{(D')^2} \tag{10.19}$$

By using Eqs. (10.7) through (10.12), the canonical circuit elements appear as:

$$E = \frac{-V}{D^2} \tag{10.20}$$

$$f_1 = 1 - \frac{sL_eD}{R} \tag{10.21}$$

$$J = -\frac{V}{R(D')^2} \tag{10.22}$$

$$f_2 = 1 \tag{10.23}$$

$$H_e = \frac{1}{1 + s(L_e/R) + s^2 L_e C} \tag{10.24}$$

Looking at the expression for H_e in Eq. (10.24), it is very evident that this transfer function is characteristic of a single-section LC low-pass filter. Finally, the continuous mode small-signal model for the buck-boost converter is constructed, using the element values given above. The complete model is shown in Fig. 10.3, where the corresponding "negative" values of E and J are represented

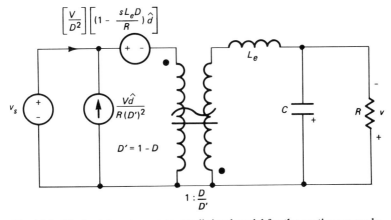

Fig. 10.3. The buck-boost converter small-signal model for the continuous mode.

by voltage and current generators of reverse polarity. In a similar fashion, the "negative" nature of the value for M is represented by inverting the winding relationships of the transformer within the model.

In Chapter 7, we demonstrated that the buck-boost SPC could be derived from a cascade connection of a basic buck converter followed by a basic boost converter, if an inversion of the output voltage polarity was acceptable. Table 10.1 gives a comparison of the element values for the small-signal canonical models of the buck, boost, and buck-boost converters. Notice that the element properties of the buck and boost converters are clearly related to those of the buck-boost converter. This observation is another confirmation of Key Idea #3 given in Chapter 1.

In an analogous manner, we can use a canonical model to represent the small-signal characteristics of an SPC operating in the discontinuous mode. The general form of this model is shown in Fig. 10.4. The procedure to develop element values for the discontinuous model follows in much the same manner as that given earlier for the continuous mode.

Table 10.2 gives the discontinuous model element values for the buck, boost, and buck-boost converters. Note also that the values of j_1 and j_2 in the case of the buck-boost are dependent on output DC voltage *magnitude* only.

TABLE 10.1.   Comparison of Buck, Boost, and Buck-Boost Model Element Values.

Element	Buck	Boost	Buck-Boost
M	D	$\dfrac{1}{D'}$	$-\dfrac{D}{D'}$
E	$\dfrac{V}{D^2}$	V	$-\dfrac{V}{D^2}$
J	$\dfrac{V}{R}$	$\dfrac{V}{R(D')^2}$	$-\dfrac{V}{R(D')^2}$
f_1	1	$1-\dfrac{sL_e}{R}$	$1-\dfrac{sL_eD}{R}$
f_2	1	1	1
L_e	L	$\dfrac{L}{(D')^2}$	$\dfrac{L}{(D')^2}$

$D' = 1 - D$

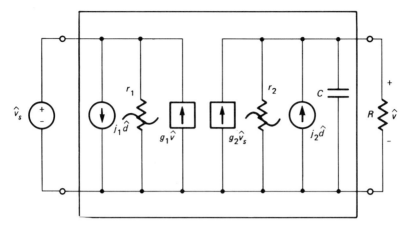

Fig. 10.4. The small-signal canonical model for the discontinuous mode.

10.2 MODELS FOR COMPLEX CONVERTERS

For most converter circuits, it is not necessary to go through the lengthy and somewhat complex mathematical derivations that were performed in the examples of the last section. The equivalent small-signal circuit model can usually be deduced from a knowledge of its derivative basic converter along with a little bit of manipulation of the canonical model to add any new circuit features of the SPC under analysis.

In Chapter 5, it was shown that the quasi-squarewave converter was derived by inserting a DC transformer into a basic buck converter (Figs. 5.12, 5.13). To derive the idealized small-signal canonical model for the quasi-squarewave converter, it is only necessary to add the DC transfer function to the canonical model for the buck converter. This addition is shown in Fig. 10.5A, where N represents the turns ratio of the DC transformer. This model can be simplified further by simply merging the two transformer functions into one element as shown in Fig. 10.5B.

The presence of a second transformer element in Fig. 10.5A will modify the E and J factors of the model by its turns ratio, N. These changes are needed as this second transformer will reflect the model's DC output voltage and load resistance by factors of $(1/N)$ and $(1/N)^2$, respectively.

Table 10.3 gives comparisons between model element values for the buck converter, its quasi-squarewave offspring, plus those for a two-output SPC of the same kind. For a quasi-squarewave converter with more than one output, the model development process is still basically the same but one additional step must be added. Fig. 10.6A shows an example of the development of a small-signal model for a dual-output quasi-squarewave SPC. The first step in reducing

TABLE 10.2. Element Definitions for Model of Fig. 10.4.

SPC Type	M	j_1	r_1	g_1	j_2	r_2	g_2
Buck	$\dfrac{2}{1+\sqrt{1+8\tau_L/D^2}}$	$\dfrac{2V}{R}$	$\left[\dfrac{1-M}{M^2}\right]R$	$\left[\dfrac{M^2}{1-M}\right]\dfrac{1}{R}$	$\dfrac{V}{RM}\sqrt{\dfrac{2(1-M)}{\tau_L}}$	$(1-M)R$	$\dfrac{M(2-M)}{(1-M)}R$
Boost	$\dfrac{1+\sqrt{1+2D^2/\tau_L}}{2}$	$\dfrac{V}{R}\sqrt{\dfrac{2M}{\tau_L(M-1)}}$	$\left[\dfrac{M-1}{M^3}\right]R$	$\left[\dfrac{M}{M-1}\right]\dfrac{1}{R}$	$\dfrac{V}{RM}\sqrt{\dfrac{2M}{\tau_L(M-1)}}$	$\left[\dfrac{M-1}{M}\right]R$	$\dfrac{M(2M-1)}{(M-1)}R$
Buck-Boost	$\dfrac{D}{\sqrt{2\tau_L}}$	$\dfrac{\sqrt{2}|V|}{R\sqrt{\tau_L}}$	$\dfrac{R}{M^2}$	0	$\dfrac{|V|}{RM}\sqrt{\dfrac{2}{\tau_L}}$	R	$\dfrac{2M}{R}$

$\tau_L = L/RT_S = Lf_S/R$
D = switch conduction duty cycle

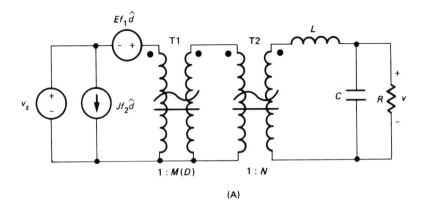

(A)

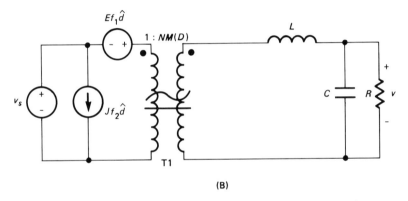

(B)

Fig. 10.5. Derivation of the small-signal continuous-mode model for the quasi-squarewave converter.

TABLE 10.3. Buck-Derived Model Element Comparisons.

Element	Basic Buck	Single-Output Quasi-Squarewave	Dual-Output Quasi-Squarewave
M	D	nD	n_1D
E	V/D^2	V/nD^2	V_1/n_1D^2
J	V/R	nV/R	$(n_1V_1)/(R1\|(n_1/n_2)^2R2)$
f_1	1	1	1
f_2	1	1	1

n, n_1, n_2 transformer turns ratios
$\|$ in parallel with.

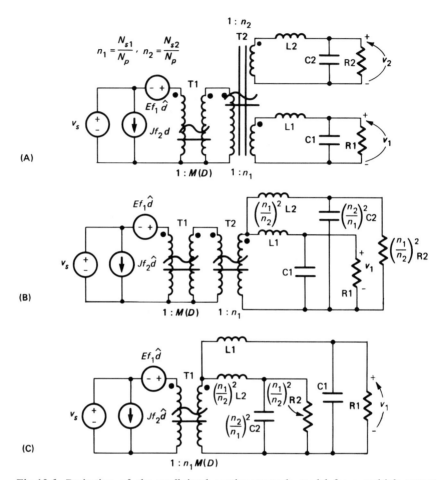

Fig. 10.6. Derivation of the small-signal continuous-mode model for a multiple-output quasi-squarewave converter.

the model of Fig. 10.6A is the transformation of the circuit elements on winding N_{s2} to winding N_{s1}, as is done in Fig. 10.6B. We can then combine the two transformer functions as was done in the last example in order to arrive at the final model in Fig. 10.6C. Normally, the reason for deriving the small-signal model of an SPC is to facilitate analyses that examine the stability effects of adding a voltage control feedback system associated with one of the outputs. To represent the power converter model as an element in an overall control system, Fig. 10.6C can be redrawn as shown in Fig. 10.7. In particular, notice here that the equivalent low-pass filter network is quite different from that of a single-output case as the low-pass filter elements associated with winding N_{s2} reflect

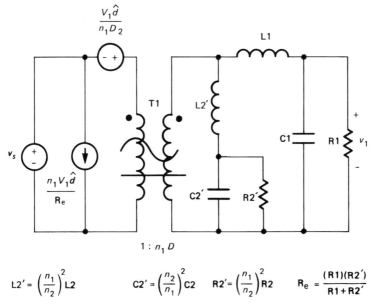

$$L2' = \left(\frac{n_1}{n_2}\right)^2 L2 \qquad C2' = \left(\frac{n_2}{n_1}\right)^2 C2 \qquad R2' = \left(\frac{n_1}{n_2}\right)^2 R2 \qquad R_e = \frac{(R1)(R2')}{R1+R2'}$$

Fig. 10.7. Multiple-output quasi-squarewave converter model reorganized.

as shunt elements at the input of the low-pass filter connected to winding N_{s1}! The control-to-output transfer function of this multi-output SPC can be quite different from that of its parent, the buck converter, when a nonzero input impedance is present. This transfer function change must be taken into account if proper stability of output voltage control is to be achieved. This example points out once again the great power and utility of the canonical small-signal models.

The complexity of the low-pass filter in this example is not characteristic of all small-signal models of multi-output buck-derived circuits. As an example, the small-signal model for the dual-output converter of Fig. 5.6B in Chapter 5 is shown in Fig. 10.8A. Again, the number of windings on T2 are reduced by transforming the components of winding N_{s2} to positions associated with winding N_{s1}. Next, we move inductor L to the secondary of T2 and then combine the functions of T1 and T2. The final model of this SPC is shown in Fig. 10.8B. In this model, notice that its low-pass filter retains the simple single-section LC characteristic of the basic buck converter model.

Continuing the simple process of SPC model development, let us turn now to examples of boost-derived converters. It was demonstrated in Chapter 6 (Fig. 6.4) that the overlapping-switch-conduction family of SPCs were derived by inserting a DC transformer into a basic boost converter. Knowing this, we can immediately draw the model of Fig. 10.9A. This network can be further simplified as that shown in Fig. 10.9B by translating inductor L to the secondary

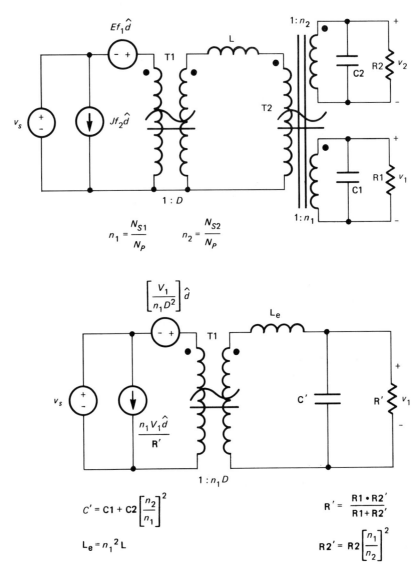

Fig. 10.8. Derivation of the small-signal continuous-mode model for a multiple output Cronin converter.

side of T2 and then combining the function of T1 and T2 into one transformer element.

When a converter circuit uses a tapped inductor or transformer, the derivation of its small-signal model can be somewhat more involved. For example, if a

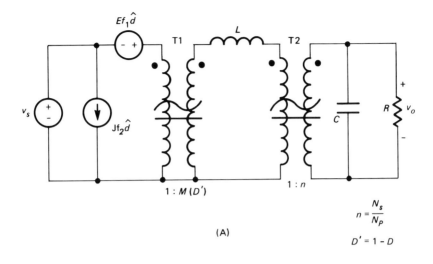

(A)

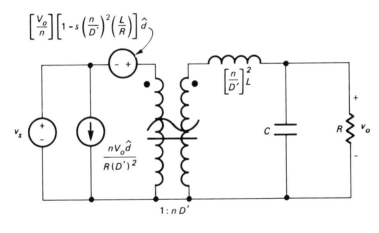

(B)

Fig. 10.9. The small-signal continuous-mode model for the Clark circuit.

tapped inductor is used in a buck converter as shown in Fig. 10.10A, the corresponding small-signal model will change as indicated in Fig. 10.10B, where:

$$L_e = \frac{L}{[D' + D(N_1/N_2)]^2}. \qquad (10.25)$$

In this example, in exchange for reducing the voltage across S1, the input voltage generator now has a zero dependent on R, L_e, and D. Also, the poles

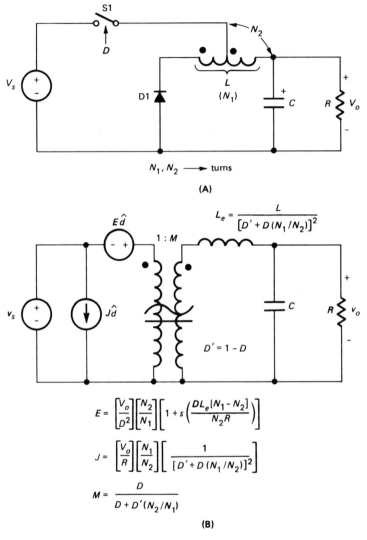

Fig. 10.10. The small-signal continuous-mode model for a tapped-inductor buck converter.

of the model's low-pass filter will now vary as a function of D because of the dependence of L_e on switch duty cycle. This change in the cutoff frequency (Δf_c) is:

$$\Delta f_c \propto \frac{1}{\sqrt{\Delta L_e}}. \tag{10.26}$$

Therefore, the effect of output filter cutoff frequency change or that of a zero in $E\hat{d}$ may not be a problem if the tap ratio (N_1/N_2) is not much greater than unity. Also, if the range of duty cycle D is not too great, changes in L_e will be small and could be neglected. If either of these two situations are not the case, reducing the switch voltage stress by tapping the inductor is probably a poor trade for increased control stabilization problems.

A second example of more complex model values is the tapped-transformer buck-derived converter shown in Fig. 10.11A. The small-signal model for this SPC variation is given in Fig. 10.11B, which shows quite clearly the effect of the variable turns-ratio of the transformer, T1.

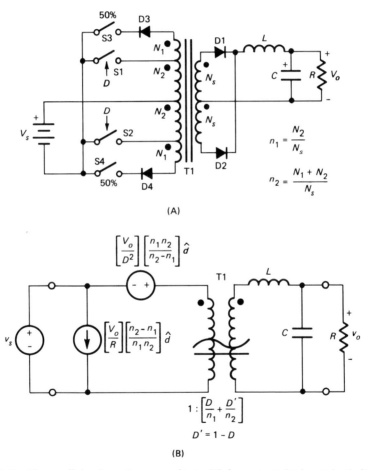

(A)

(B)

Fig. 10.11. The small-signal continuous-mode model for a tapped-primary buck-derived converter.

10.3. CLOSING THE LOOP

Normally, the output from a power converter is required to be regulated against changes in its load and input voltage. Usually this control is accomplished by auxiliary electronics that compare the output voltage to that of a constant reference voltage. These electronic networks will in turn produce a control voltage (v_c) that is a function of this comparison. The control voltage magnitude is then converted to a value of d using a pulse-width-molulator (PWM) circuit. The PWM output is then employed to drive the switches of the power converter. This negative feedback arrangement is shown schematically in Fig. 10.12. There are many ways in which a PWM can be realized, but its general function can be rather succinctly stated as follows:

$$d = \left(\frac{f_m}{V_m}\right) v_c \qquad (10.27)$$

where V_m is the DC range of v_c required to vary the value of D from 0 to 1 and f_m is the generalized AC transfer function characteristic of the particular PWM under consideration. For the purposes of this discussion, we will leave the small-signal PWM control-to-duty cycle gain function of Eq. (10.27) in the form shown.

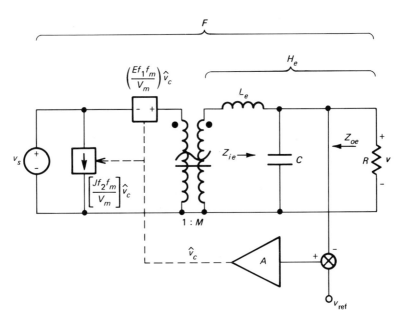

Fig. 10.12. The continuous-mode canonical model combined with an output voltage control system.

In the model of Fig. 10.12, the small-signal portion of the PWM gain function has been merged with the voltage and current generator element values. Also notice that these generator symbols are now represented by "squares," indicating that they are *dependent* sources controlled by the small-signal value of $\hat{v}_c$. The methods of linear circuit analysis can now be applied directly to the model of Fig. 10.12 to derive the system's electrical characteristics in a straightforward manner. For example, the open-loop control loop gain (T) can be found by inspection of Fig. 10.12 as:

$$T = \frac{E f_1 f_m A M H_e}{V_m}.$$

(10.28)

Notice that T does not contain a term for the dependent current generator of the canonical model $(J f_2)$. In this example, the model is driven from a zero-impedance voltage source (v_s). Therefore, this current generator plays no effective role in determining the various transfer functions of the system model. Usually, this is not the case, as we shall soon see when an input filter network is added between the voltage source and the input of the SPC model. From elementary control theory we know that:

$$Z_0 = \frac{Z_{00}}{1 + T}$$

(10.29)

$$F = \frac{\hat{v}_o}{\hat{v}_s} = \frac{M H_e}{1 + T}.$$

(10.30)

Finally, with a modest algebraic exercise, the closed-loop input admittance $(1/Z_i)$ can be found to be:

$$\frac{1}{Z_i} = \left(\frac{1}{1 + T}\right) \left(\frac{M^2}{Z_{ie}}\right) - \left(\frac{T}{1 + T}\right) \left(\frac{M^2}{R f_1}\right).$$

(10.31)

Looking at Eq. (10.31), we see that the input admittance of the system has two terms, one of which is negative! In any conventional control loop, the value of T will be very large at low frequencies, growing smaller as frequency is increased. Also, at low signal frequencies, the factor f_1 in our model will be close to unity. Therefore:

$$Z_i \approx - \frac{R}{M^2}.$$

(10.32)

Under such conditions, the input impedance of the SPC is negative! This result is not very surprising when one considers a basic characteristic of any SPC regulator. The output voltage (V_o) is maintained constant; therefore, for a fixed value of R, the output power must also be constant. If we assume a highly efficient converter, then:

$$P_i \approx P_o = V_s I_s . \tag{10.33}$$

Since P_i is constant, I_s must decrease as V_s increases and vice versa. Therefore, the incremental input resistance (R_i) of the regulator system must be *negative*. At low frequencies, then:

$$Z_i \approx R_i = - \frac{R}{M^2} . \tag{10.34}$$

As signal frequency is increased, the real part of the complex impedance Z_i (Re $[Z_i]$), will at some point become positive in value; however, there is a substantial range of frequencies over which R_i is negative. This latter situation can be a problem if an input LC filter is used between the source voltage and the SPC regulator. When an LC filter is terminated in a negative resistance, self-oscillation can occur if the total effective damping resistance (i.e., the termination resistance plus any filter-damping resistances) becomes negative. Very few SPCs are built without an input filter for conducted EMI control. Therefore, this negative impedance problem can be very serious unless means are taken to circumvent self-oscillation due to the marriage of the input filter and the SPC.

To examine this particular problem in more detail, let us add an input filter to our system model, as shown in Fig. 10.13. Here, H_s is the *unloaded* voltage transfer function for the input filter network. To somewhat simplify our analysis, the voltage source and the input filter can be transformed into a Thevenin equivalent network, as shown in Fig. 10.14. Impedance Z_s represents the parallel combination of C1 and the series network consisting of R1 and L1. By applying some muscle to the crank of our algebra machine, we can derive the following new relationships, which now contain the effect of the input filter network:

$$T' = T \left[\frac{1 - \dfrac{M^2 f_1 f_2 Z_s}{R}}{1 + M^2 (Z_s / Z_{ie})} \right] \tag{10.35}$$

$$F' = F \left[\frac{1}{1 + M^2 (Z_s / Z_{ie})} \right] \left[\frac{1 + T}{1 + T'} \right] \tag{10.36}$$

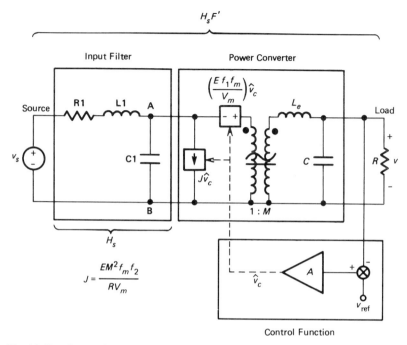

Fig. 10.13. The continuous-mode small-signal model for a complete converter circuit.

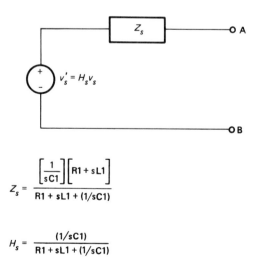

$$Z_s = \frac{\left[\dfrac{1}{sC1}\right]\left[R1 + sL1\right]}{R1 + sL1 + (1/sC1)}$$

$$H_s = \frac{(1/sC1)}{R1 + sL1 + (1/sC1)}$$

Fig. 10.14. Equivalent circuit representing the input source and low-pass filter.

$$Z_0' = Z_0 \left[\frac{1 + M^2(Z_s/Z_a)}{1 + M^2(Z_s/Z_{ie})} \right]. \tag{10.37}$$

Here, Z_a is the value for Z_{ie} when the output of the system is short-circuited (i.e., $R = 0$). Therefore, in this particular example:

$$Z_a = sL_e. \tag{10.38}$$

The values of T, F, and Z_0 in Eqs. (10.35) through (10.37) are those obtained from the expressions of Eqs. (10.28) through (10.30), respectively (i.e., *without* an input filter present).

Equations (10.35) through (10.38) express a number of important features now evident in the SPC. First of all, the inclusion of the input filter network alters *all* of the essential SPC circuit properties. Unfortunately, these changes are for the worse. For example, output impedance (Z_0) and line-to-output transmission (F) are increased, while loop gain (T) is decreased. Looking at Eq. (10.35), we see that there is a negative term in the numerator of this equation. Therefore, the sign of the system's loop gain can change if:

$$\frac{M^2 f_1 f_2 Z_s}{R} > 1. \tag{10.39}$$

If the condition of Eq. (10.39) is satisfied, the regulatory system will probably become unstable and oscillate. Even if oscillation does not occur output transient response to changes in load and input voltage will be degraded to the point where serious "ringing" of output voltage will occur.

To assure stability of the SPC system with an input filter, two *necessary* and *more than sufficient* conditions must be met:

$$\frac{M^2 f_1 f_2 Z_s}{R} \ll 1 \tag{10.40}$$

$$\frac{M^2 Z_s}{Z_{ie}} \ll 1. \tag{10.41}$$

Placing the above conditions into the earlier expressions for T', Z_o', and F', we find that:

$$T' \approx T \tag{10.42}$$

$$Z_o' \approx Z_o \tag{10.43}$$

$$F' \approx F. \tag{10.44}$$

It is apparent from these relationships that the input filter will not materially affect converter system performance if they are satisfied. The key factor is to make Z_s as small as possible (large C on the output). In some SPC regulator applications, such as high power operation with very restrictive EMI and weight requirements, it may not be practical to meet the criterion of Eqs. (10.40) and (10.41). In such instances, detailed modeling of the complete converter system using computer-aided analysis will probably be necessary to quantitatively evaluate the degradation in performance due to the presence of an input filter. Fortunately, the models we have developed in this chapter are well suited for such analyses. For further discussion of the input filter/SPC interaction problem, the reader is referred to Refs. 7 and 8 in the bibliography of this text.

11. Comparative Techniques for SPC Selection

In the preceding chapters of this book, we have presented a large number of different SPC circuits. In addition, means by which even more SPC circuit variations can be created have been demonstrated. This overwhelming wealth of converter topologies is almost too much of a good thing because it creates a very basic problem, namely, how do we choose the best converter approach for a given application from the multitudes of options available to us?

Each individual circuit has many characteristics which must be considered for a particular application. Some of the more common application characteristics which must be compared are listed below:

1. Required control-loop and input-to-output dynamic transfer functions,
2. Component stress levels,
3. Circuit complexity (component count),
4. Efficiency of power conversion,
5. Weight of the converter system,
6. Volume of the converter system,
7. System filtering requirements for input and outputs,
8. Energy storage required (both capacitive and inductive),
9. Producibility and maintainability,
10. Cost of the conversion system,
11. Effect of nonideal components on desired performance, and
12. Unusual electrical or packaging needs.

One means for simplifying the selection problem is to arrange the circuits into groups that exhibit common properties. Once this is done, we can zero in on the general characteristics of the SPC quickly and then turn to considering the differences between a small number of circuit variations in greater detail.

As a practical matter, whatever process we choose to use must converge on the best choice relatively quickly. Engineering time is usually limited and expensive and designers can rarely afford to spend any great length of it trying to select the one most perfect SPC for a particular power conversion application need.

In this chapter we will examine means for categorizing and comparing SPC circuits. While these methods are helpful, keep in mind that they do not guarantee an optimum choice. You, the designer, will have to exercise a good deal of judgment to steer the selection process in the right direction to achieve an optimum solution. Like most engineering problems, there are often several solutions that are very nearly equal and viable with the final choice usually made as a result of personal preference.

11.1. FAMILIES OF CONVERTER CIRCUITS

The most obvious category scheme that comes first to mind is the grouping of SPC circuits on the basis of their basic parent converter (i.e., buck, boost, type of DC transformer, etc). As we have seen, the properties of the basic constituents are, for the most part, preserved in related converters. From a knowledge of basic SPCs, we can therefore predict in advance most of the conversion properties of any derivative.

We have chosen to group SPC circuits by the following criterion; first, the basic converter or combination thereof; second, by the choice of DC transformer if one is used and, finally, by whether or not the transformer is voltage-fed or current-fed.

This scheme makes a great deal of sense since, from the basic properties of a converter, we can quickly determine electrical transfer functions, switch duty cycles, component currents, etc. Then, from the DC transformer, we can rapidly determine the switch voltage and current stresses.

An example of this ordering scheme is shown in Fig. 11.1. Here we have developed a family tree for relating the circuits developed in earlier chapters of this text. The numbers shown in each portion of the tree relate to the figure number of this book.

The grouping begins with the buck and boost converters, which are dual in their electrical structures. From these two basic circuits, three major branches are delineated—those converters derived from the buck converter, those derived from the boost converter, and those derived from a combination of both buck and boost converters. Each main branch of the tree is further divided by the type of DC transformer (or the lack thereof) chosen. Notice that a hybrid branch has been included in Fig. 11.1 for the DC transformer category. The hybrid branch takes into account the possibility of a multiple input converter (Fig. 8.18) in which one input uses a different switch connection than the other input.

The next level of division is by the type of source at the input to the DC transformer (current or voltage-fed). By now, our family tree is looking pretty busy. Therefore, we have omitted cross linkage to indicate the duality relationships between some members of the buck- and boost-derived families. While de-

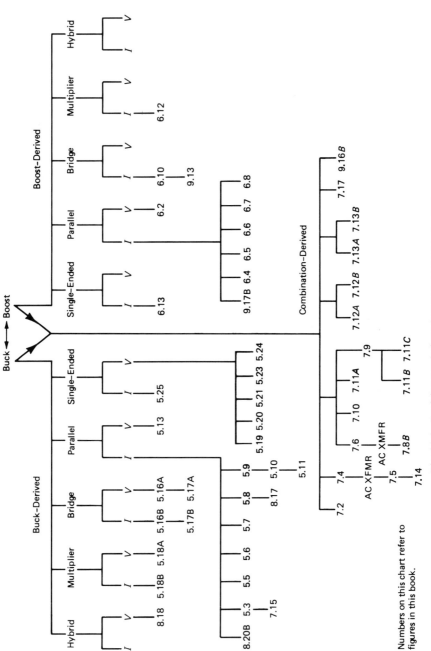

Fig. 11.1. Evolving a switchmode power converter family tree.

sirable, addition of such duality relationships would render Fig. 11.1 completely illegible and perhaps unusable for some readers.

By way of an example of the use of this family tree, let us look at what it tells us about the quasi-squarewave converter of Fig. 5.13. In Fig. 11.1, we see immediately that it is a buck-derived converter and uses a voltage-fed parallel transformer. From these facts we can immediately infer that:

1. The input current is pulsating in nature.
2. The output current is nonpulsating in nature.
3. The input-to-output and the control-loop transfer functions have simple low-pass characteristics with fixed left-half-plane poles and zeros.
4. The ideal peak voltage that each of the converter switches will see when in an OFF state will be twice the supply voltage, V_s.
5. The average current of either switch will be equal to one-half of the average input current.
6. Because the transformer is voltage-fed, the switches may be subjected to current spikes should conduction overlap occur.

Thus, at a glance, we are able to define most of the important properties of a selected circuit. It is for this reason that this method of converter circuit grouping is so useful. By way of comparison, we could repeat the selection process with another circuit and then cancel out common characteristics leaving only the differences to compare if they are important for the application under consideration.

Of course, it would be impractical (unless you are willing to give up one wall of your office!) to include all possible circuits on a single diagram. So this type of diagram usually resides in the mind of the designer along with the information of the derivation of each circuit. When actually selecting a circuit, only portions of the family tree would need to be drawn by a designer as aids in converter selection.

11.2. QUANTITATIVE COMPARISONS

The qualitative comparison aided by family tree grouping is very helpful in narrowing the field of search for the right converter structure. However, sooner or later, a quantitative comparison must be conducted to determine actual circuit characteristics and the obtainable performance of each candidate circuit.

As a typical example of quantitative analysis at an intermediate level, we will examine a comparison between the buck-derived parallel quasi-squarewave converter of Fig. 5.13 and the boost-derived parallel overlapping-conduction converter as shown in Fig. 11.2. From the assumptions listed in Table 11.1 and the equations given in Tables 2.2A and 3.2A, we can derive the worst-case values for

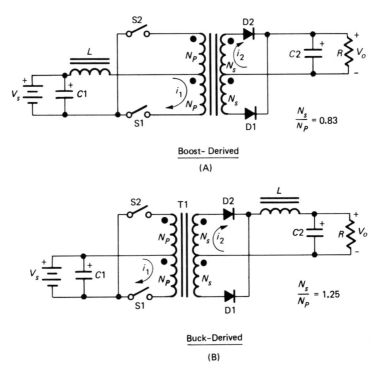

Fig. 11.2. Buck- and boost-derived circuits for comparison.

the circuit currents and voltages. The results of the derivations are delineated in Table 11.2.

From Table 11.2, we can make the following four observations:

1. If we assume that the size of L is proportional to the peak energy stored in it, then the buck inductor is physically larger by a factor of 1.5 over the inductor of the boost converter. Looking at $I_{L\,(\text{rms})}$, we see that this current is larger in the boost converter, but since the inductance is lower in value, the boost in-

TABLE 11.1. Circuit Parameters for
Performance Comparisons.

$$V_s = 20 - 30 \text{ V}$$
$$V_o = 25 \text{ V}$$
$$R = 1\,\Omega$$
$$f_s = 50 \text{ kHz}$$
$$\tau_L = 2\,\tau_{LC}$$

Switch and diode voltage drop = 0

TABLE 11.2.   Continuous-Mode Buck and Boost Converter Comparison (see Table 11.1).

	Buck	Boost	Ratio (Ratio^2)
I_b	46.9 A	46.7 A	1.0
$I_{C1(\text{rms})}$	16.5 A	2.6 A	–
$I_{C2(\text{rms})}$	7.2 A	19.8 A	–
$(I_{C1} + I_{C2})_{(\text{rms})}$	23.7 A	22.4 A	1.06 (1.12)
$I_{T(\text{rms})}$	22.1 A	21.1 A	1.05 (1.10)
$V_{T(\text{max})}$	60 V	60 V	1.0
L	3.3 μH	2.2 μH	1.5
$I_{L(\text{rms})}$	26.1 A	31.2 A	0.84 (0.71)
Peak Energy $(\frac{1}{2}I_b^2 L)$	3.6×10^{-3} J	2.4×10^{-3} J	1.5
Duty cycle range (S1 or S2)	0.33 to 0.5 (1.5 : 1)	0.5 to 0.665 (1.33 : 1)	1.13
$I_{1(\text{rms})}$	22.1 A	21.1 A	1.05 (1.10)
$I_{2(\text{rms})}$	18.4 A	22.6 A	0.81 (0.66)
$I'_{2(\text{rms})}$ (I_2 reflected to primary)	23.0 A	18.8 A	1.22 (1.50)
$(I_1 + I'_2)_{(\text{rms})}$	45.1 A	39.9 A	1.13 (1.28)

ductor will have fewer turns and its copper loss will be approximately the same as the inductor of the buck converter.

2. The total rms capacitor current is slightly larger in the buck converter. It is common design practice for the capacitor size to be determined more by current requirements than by the capacitance so that the buck converter requires a slightly greater volume of capacitors.

3. The total rms winding current reflected to the primary $(I_1 + I'_2)$ is greater for the buck; but, since the primary winding in the buck will have a factor of 0.67 times the number of turns of the primary in the boost converter, the total copper loss in the buck transformer for the same wire size will be 85% that of the boost. For the same losses, the boost transformer will be slightly larger.

4. $I_{T(\text{rms})}$ is 5% larger in the buck converter. If MOSFET switches are used where their conduction losses are proportional to $[I_{T(\text{rms})}]^2$ and their ON re-

sistances have a positive temperature coefficient, then the conduction losses could well be 20–30% higher in the buck converter depending on the thermal impedances of heat sinks used for the MOSFETs. The buck converter will need larger switches for the same power level.

From the preceding observation, it would appear from Table 11.2 that the boost circuit will be somewhat smaller or more efficient than the buck and therefore must be the logical choice. Unfortunately, this simple example has ignored some other practical considerations that make the choice not so clear cut. For example:

1. If an EMI specification such as MIL-STD-461A is imposed on the input to the converter, and the output noise specification is substantially wider in tolerance (a fairly common requirement), then the input filter for the buck converter will be much larger, since the input current is pulsating. On the other hand, if a low-noise EMI specification is required of the output, then the total buck converter system may very well be smaller in volume and weight.

Consider also that increasing L in the boost converter reduced $I_{C1\,(rms)}$, as opposed to the buck converter where increasing L above $2L_C$ doesn't change the value of $I_{C1\,(rms)}$ to a great degree. The dual argument applies to the output filter.

2. The control-to-output characteristic for the buck converter is very well behaved, but the boost converter has moving poles and a right-half-plane zero. As a result, efforts to provide unconditional stability usually result in the boost converter control loop having less bandwidth and, therefore, a slower transient response. The high-frequency output impedance of the boost converter will be lower since the output capacitor is so much larger (2.8:1), which compensates somewhat for the smaller bandwidth.

3. The boost-derived converter is current-fed by nature, as was pointed out in Chapter 4. The circuit will have much less of a problem dealing with transformer saturation and most other nonideal component behavior. An exception, of course, is the problem of high voltage spikes caused by transformer leakage inductances.

4. If overload, turn-on, and turn-off output voltage-ramping and inrush-current controls are required, then the buck converter is a much better choice. To exhibit these often necessary properties, the boost converter must be made more complex in topology in order to emulate a buck converter when called upon to do so.

11.3 THE PRACTICAL COMPARISON PROCESS

The preceding discussion has shown that both quantitative and qualitative considerations must enter into any decision process by which a particular SPC circuit is chosen for a given application.

In practice, the decision process usually begins rather subjectively (with personal bias and experience playing an important role) with qualitative considerations using the family tree approach to narrow the number of alternatives. The process then becomes more objective as quantitative comparisons are made. The final converter choice is then made, based on a detailed quantitative comparison of a very few circuits. This final stage of comparison often requires the execution of very detailed circuit designs each using one candidate converter. Since this final step is very time-consuming, it is rarely possible to do such detailed comparisons between more than two or three circuit approaches. For this reason, the judgment (and intuition) exercised by a designer at the beginning of the selection process will be the major factor in determining final converter choice and its degree of acceptability. Like most human endeavors, such judgment comes with experience. Therefore, the inexperienced designer would be well advised to spend a good deal of effort at an intermediate level of quantitative comparison of many different converter approaches before choosing final candidates. Repeating this comparison process for several converter designs will bring familiarity and speed of selection for future applications.

12. Converters with Integrated Magnetics

The continuing and seemingly never-ending quest by product designers to develop new techniques to integrate electronic circuit functions has motivated power electronics engineers to find ways whereby the physical size and parts count of SPCs can also be reduced. One part of these investigative efforts that has shown definite promise in recent years is the integration of the magnetic components of an SPC. Such methods provide for the combination, or "lumping," of inductors and transformers into *single* physical assemblies with little or no compromise in conversion characteristics.

As we have seen in past chapters, SPC inductors and transformers perform a variety of essential roles and are key to the dynamic power transfer and storage processes of these circuits. Unfortunately, and with few exceptions, inductors and transformers are also the major contributors to the total cost, weight, and size of a converter system. Therefore, it is not very surprising that SPC designers often go to considerable extremes to minimize the number and size of magnetic components required in a converter, and not uncommon that such extremes include the selection of a less-than-optimum SPC circuit approach that, in turn, calls for one or more undesirable concessions in electrical performance.

SPC topologies which inherently adapt well to the integration process of magnetic functions are very attractive to designers for reasons such as those just mentioned. Also, magnetic integration, if properly executed in the design of these converters, can bring added benefits in electrical performance, such as reduced stress on other SPC components or lower ripple currents on input and output lines. The Ćuk converter variations discussed in Chapter 7 (Fig. 7.11) and the multi-winding output inductor assembly proposed in Chapter 8 (Fig. 8.4) are excellent examples where integration of magnetic functions will not only reduce component count, but also will enhance the operational features of these converters. The flyback converter of Chapter 7 (Fig. 7.4B) is another example, wherein its singular magnetic serves as the primary energy storage component of this SPC as well as a means for providing electrical isolation between all input and output terminals.

In this chapter, we will explore the unusual subject of integrating magnetic circuit elements of SPCs. In our exploration, we will take a more detailed analytical look at the single-magnetic Ćuk converter of Fig. 7.11C to better understand how it can be designed to have *both* zero input and zero output current ripple. Because the output section of a Ćuk converter is essentially that of a buck SPC, this look will also provide insight into the design principles of a multi-winding inductor assembly for use in the quasi-squarewave converters of Figs. 8.4B and 8.5A. Finally, magnetic design methods whereby the inductors and transformers of many other SPC circuits can be blended together will then be examined, followed by the evolution of new and unique converter circuits using these integration methods.

12.1. MAGNETIC CIRCUIT ANALYSIS—A REVIEW OF CONCEPTS

Magnetic circuit concepts are subjects usually found in basic electrical engineering textbooks dealing with topics related to electrostatic field theory. For the most part, these important concepts are usually presented in such books with little or no illustration of application in the practical design of inductors and transformers for SPCs.

While a complete review of the fundamentals of magnetic circuits is beyond the scope of this book, a brief review of those important to the proper understanding of ferromagnetic circuits is appropriate here, as a prelude to later discussions of SPC inductor and transformer integration methods. For those readers interested in a more thorough exposure, Ref. 9 cited in the bibilography of this text is highly recommended.

Recall that the similarity between Kirchhoff's voltage and current laws for linear electric circuits and Ampere's circuit laws related to magnetomotive force and flux continuity in *linear* magnetic circuits permit the use of electric circuit analogs for analysis purposes. These analysis methods make voltage analogous to magnetic potential, F, current analogous to magnetic flux, ϕ, and electrical resistance analogous to magnetic reluctance, $\mathcal{R}$. It follows also that any electrical circuit resulting from application of these analogous relationships will have the same topology as that of its magnetic circuit counterpart. Furthermore, because it is linear, this circuit model can be manipulated into even more useful forms by the use of the standard duality relationships explained in detail in Chapter 9. Conversion by duality will then provide a final model relating reluctance values to inductances, flux linkages in windings to voltages, and flux magnitudes in magnetic paths into currents.

The familiar relationship between the voltage appearing across an inductor L and the rate of change of current through it is simply:

$$v = L \frac{di}{dt} \tag{12.1}$$

in accordance with Faraday's classic law of electromagnetic induction. This law, when applied to a single coil of wire with N turns, can also be restated as:

$$v = \frac{d\lambda}{dt} = N\frac{d\phi}{dt} \qquad (12.2)$$

where λ is the measure of *flux linkage* within the coil produced by *self-induction*. In those instances where multiple coils of wire have common magnetic paths, the total flux linkage in any one coil would be the sum of that produced by self-induction plus those produced by *mutual induction*, i.e., caused by dynamic changes in the currents of the remaining coils. For the case of a single coil, its *self-inductance* can be determined by the equality of Eqs.(12.1) and (12.2) as:

$$L = N\frac{d\phi}{di}. \qquad (12.3)$$

Magnetomotive force (F) of an excited coil of wire is defined as the product of the instantaneous current through it and the number of turns of the coil. In equation form, F can be directly related to *magnetizing force*, H, and its magnetic path length, l, as:

$$F = Ni = Hl. \qquad (12.4)$$

Assuming that a linear relationship between ϕ and H is always maintained, the *reluctance* $(\mathcal{R})$ of a magnetic path can be defined as the ratio of a change in magnetomotive force to the corresponding change in flux value. Therefore, it follows from Eqs. (12.3) and (12.4) that:

$$\mathcal{R} = \frac{dF}{d\phi} = N\frac{di}{d\phi} = \frac{N^2}{L}. \qquad (12.5)$$

Reluctance can be expressed in terms of related magnetic path length, cross-sectional area (A_C), and the permeability (μ) of the magnetic path. Assuming that the cross-sectional area is uniform,

$$\mathcal{R} = \frac{l}{\mu A_C} = \frac{1}{\mathcal{P}} \qquad (12.6)$$

where $\mathcal{P}$ is defined as material *permeance*, the reciprocal element of reluctance.

In magnetic circuits, arrows are used to indicate the assumed directions of currents rather than polarity marks for magnetomotive forces (mmfs). The best known method for relating the direction of current to that of the flux it produces is the "right-hand rule" illustrated in Fig. 12.1A.

With the right hand positioned as shown, the flux direction will be indicated by the direction of curvature of the fingers. Note that the thumb must be aligned so as to point in the direction of the current. The fingers then point in the positive direction of flux produced by the current.

In representing transformers in electric circuit diagrams, it is customary to use *dot notation* to convey the voltage polarity relationships of windings relative to one another, as shown in Fig. 12.1B. Three basic rules are followed in establishing such relationships:

1) Voltages induced in any two windings due to changes in mutual flux will have the *same* polarity at "dot-marked" terminals.
2) If positive currents flow into the dot-marked terminals, the mmfs produced in the two windings will have *additive* polarity.

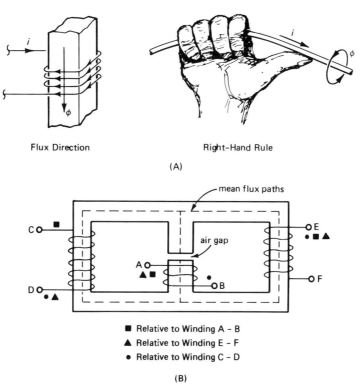

Flux Direction   Right–Hand Rule

(A)

mean flux paths

air gap

■ Relative to Winding A – B
▲ Relative to Winding E – F
● Relative to Winding C – D

(B)

Fig. 12.1. (A) Determining flux direction using the right-hand rule.
(B) Dotting of windings in a magnetic structure with multiple flux paths.

3) If one of the two windings is open-circuited, and if the current flowing into the dot-marked terminal of the other winding has a *positive* rate of change, then the voltage induced in the open winding will be *positive* at dot-marked terminal.

When a magnetic assembly has more than two windings and contains multiple magnetic material paths, then multiple "dotting" of windings will be necessary. This is easily done using the right-hand rule and the three basic relationships given above. An example of multiple dotting is shown in Fig. 12.1B. Here, three sets of "dots" are needed to express the voltage polarities relative to each of the three windings.

Finally, to complete our review, we can relate the degree of mutual induction between two windings sharing a common magnetic path by the use of the following relationship:

$$M = k\sqrt{L_{o1}L_{o2}} \tag{12.7}$$

where k is defined as the *coupling coefficient*, M is the amount of *mutual inductance* that exists between the windings, and L_{o1} and L_{o2} are the *self-inductance* (open-circuit) values of the two windings.

All of these fundamental definitions relating to magnetic circuits, when combined with the equivalent electric-circuit development methods described earlier for modeling, produce a powerful set of analysis tools for quick and accurate evaluation of any magnetic circuit system, regardless of its complexity. As demonstrations of their utility, we will now use them to develop realistic electric circuit models for an inductor and a dual-winding transformer. The reader is encouraged here to review the duality concepts of Chapter 9, for they are key to a proper understanding of these modeling examples.

12.2. EXAMPLES OF MAGNETIC CIRCUIT MODELS

Consider now the single-winding magnetic structure shown in three-dimensional form in Fig. 12.2A. Here, we have joined two C cores of equal dimensions and cross-sectional areas and have inserted small air gaps of equal length at their junctions. A winding of N turns of wire has been placed on the outer leg of one of the C cores, into which a current (i) flows.

For our analysis purposes, we will assume that flux distribution within the cores as well as across each of the two gaps is uniformly distributed, and that both cores have the same permeability, μ_m. Also, mean magnetic path lengths will be used to estimate the reluctances of each portion of the cores and that presented by the air gaps, as diagrammed in Fig. 12.2B. Using standard international MKSA (Meter-Kilogram-Second-Ampere) magnetic units, the reluctances

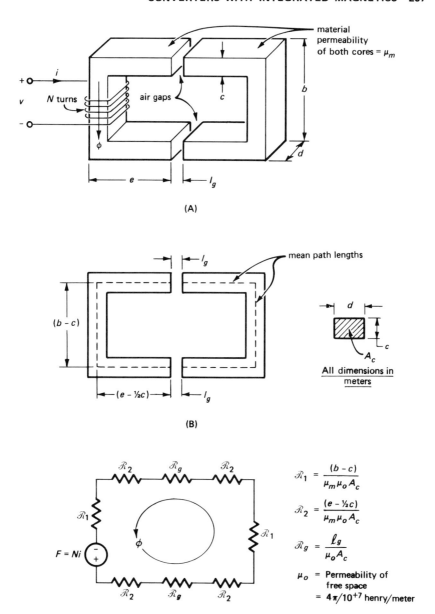

Fig. 12.2. Setting up a reluctance and mmf model for a single-winding inductor.

of each portion of the mean flux path can then be determined using the equations accompanying Fig. 12.2C. From our electric circuit analog, the flux magnitude is simply:

$$\phi = \frac{F}{\mathscr{R}_T} = \frac{Ni}{2\mathscr{R}_1 + 4\mathscr{R}_2 + 2\mathscr{R}_g} \tag{12.8}$$

where $\mathscr{R}_T$ is the sum total of all reluctances in the flux path.

Fig. 12.3A is the circuit model described by the conditions of Eq. (12.8). With an electric circuit analog established, we now determine its dual network

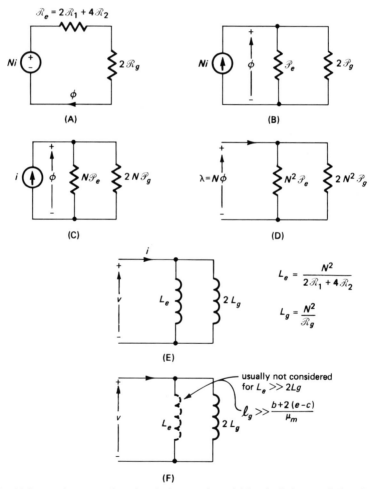

Fig. 12.3. Development of an electrical network model for the inductor of Fig. 12.2.

as is done in Fig. 12.3B. It is now necessary to scale the dual circuit by the number of turns of the winding until it is in a form where permeances can be directly related to inductances. This operation is performed in Figs. 12.3C and D. From Eqs. (12.5) and (12.6), we see that inductance is equal to permeance times the number of winding turns squared. Using this relationship, we then convert the equivalent circuit of Fig. 12.3D to its final inductive form shown in Fig. 12.3E.

Our final model clearly shows that the inductances established by the core materials are in parallel with the inductances produced by the airgaps of the composite structure. If L_e is much greater than $2L_g$, then the equivalent circuit of Fig. 12.3E reduces to the single-component network of Fig. 12.3F. In most practical inductor designs for SPCs, this is usually the case. Air gaps will be the major factors in determining their winding inductance values, and increasing air gap length will decrease the inductance of a magnetic material. However, a larger air gap will also permit more DC bias to be tolerated by the core of an inductor, a situation frequently encountered in converter circuit applications. For this reason, most SPC inductors use core structures with external air gaps in the magnetic paths, or employ cores that have inherent gaps in the composition of their magnetic material.

This simple example, evolving a practical electrical circuit model for an inductor from its magnetic structure, serves to illustrate the ease of development when electric circuit analogs are used. In our next example, we will use the same procedure to derive a practical circuit model for a transformer with leakage inductances. The magnetic circuit (taken from Ref. 9) that we use for this particular example is shown in Fig. 12.4A.

As shown, three flux paths are indicated, two of which (ϕ_2 and ϕ_1) are *not* contained within the magnetic material of the toroid. These latter paths are called *leakages* and, as we will see, produce *leakage inductances* in the final electrical model for the transformer. The remaining material flux path links both windings and is designated as the mutual flux (ϕ_m). With a voltage source applied to the primary winding (N_P), with the polarity as shown in Fig. 12.4A, the resulting current direction will produce a counter-clockwise flux in the core. Using the dot notation with the primary winding as reference, the secondary winding (N_s) will have its dot at the top of the winding, and therefore will produce a load voltage with the polarity shown. Because the secondary current direction is "out of the dot," it will produce an mmf that *opposes* that of the primary winding.

Fig. 12.4B is the equivalent magnetic circuit diagram, with its dual network shown in Fig. 12.4C. Note that ϕ_m and ϕ_1 have the same polarity in the primary winding, since they are caused by i_s. However, in the secondary winding, ϕ_m and ϕ_2 have opposite polarity, since one flux produces i_o while the other is a consequence of i_o.

With a dual magnetic network established, we can now proceed to place it

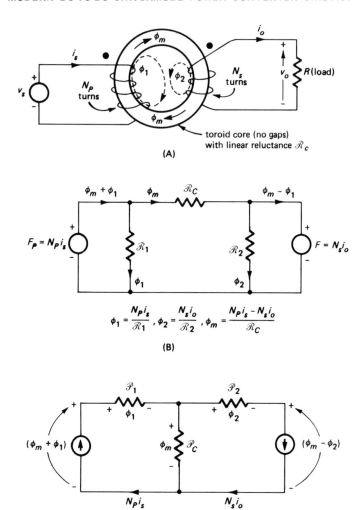

Fig. 12.4. Setting up a reluctance and mmf model for a two-winding transformer with leakage flux paths.

into an electrical form following the same steps that were used in the previous inductor example. First, we scale up the permeances of the model of Fig. 12.4C by the square of the turns of the winding that is selected as reference. In this case, the primary will be used as the reference winding. The results of this scaling operation is shown in Fig. 12.5A. Next, the scaled permeances are replaced by inductances in accord with Eq. (12.6) and flux linkages by voltage sources. This last step produces the final inductive circuit model illustrated in Fig. 12.5B.

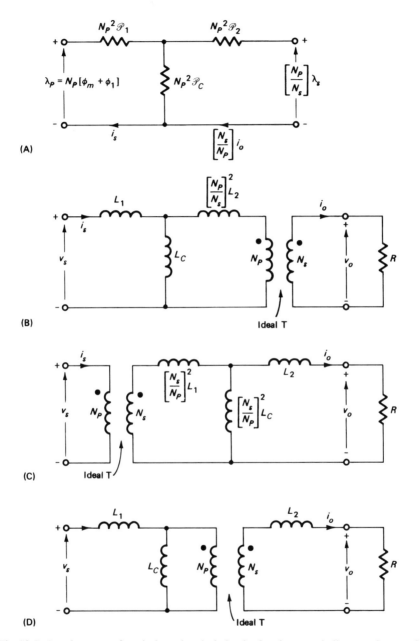

Fig. 12.5. Development of equivalent electrical circuits for the two-winding transformer of Fig. 12.4.

If one desires to reference one or more inductances of this final model to the secondary side of the transformer instead of the primary, this can be easily accomplished using impedance transformation techniques. Two alternative circuit models developed in this fashion are shown in Figs. 12.5C and D.

The last circuit model in Fig. 12.5D is a classic form used frequently to design and analyze transformers for SPCs. Here, L_1 and L_2 represent the *leakage inductances* of primary and secondary windings, respectively, while L_C is often termed the *primary magnetizing inductance* of the transformer. In practice, designers will strive to minimize the value of L_1 and L_2 in a converter transformer. This is accomplished by winding the primary and secondary in very close proximity to one another. Also, to minimize core excitation current, L_C is usually made as large as possible. Therefore, in a transformer, core structures *without* air gaps are highly desirable, although most practical designs for SPCs do have small gaps to prevent core saturation from external imbalance effects (see Section 4 of Chapter 4).

It should be pointed out here that L_C in the transformer models of Fig. 12.5 is *not* the mutual inductance (M) shared by primary and secondary windings. However, M can be easily found by writing the two nodal circuit equations that relate input and output currents of the models and isolating the common inductance term. If we choose the model of Fig. 12.5D for this analysis, the value of M is found to be

$$M = \left(\frac{N_s}{N_P}\right) L_C. \tag{12.9}$$

From Eq. (12.7), the degree of coupling between the two windings of our model can then be determined as

$$k = \frac{(N_s/N_P)\, L_C}{\sqrt{(L_1 + L_C)(L_2 + [N_s/N_P]^2\, L_C)}}. \tag{12.10}$$

Note that Eq. (12.10) correctly predicts that k will be equal to unity (perfect coupling of windings) when the two leakage inductances (L_1 and L_2) of the model are reduced to zero.

Once again, in four easy steps, an electrical circuit model that accurately represents its magnetic circuit counterpart has been found. The evolution of an electrical model by this procedure also gives much insight into the important relationships between physical core dimensions, material properties, and corresponding electrical performance. Although these two modeling examples were somewhat simplistic, the same steps of model formation can be followed in analyzing more complex and unusual magnetic circuits, as we will soon see in our discussions of SPCs with integrated magnetics.

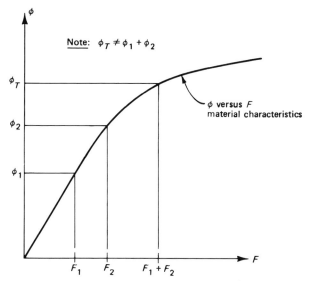

Fig. 12.6. Doubly-excited magnetic circuit operating in a nonlinear portion of the core material's characteristic curve.

Before moving into these discussions, one important modeling rule mentioned earlier needs to be re-emphasized. Always keep in mind that the electrical model analogs are based on the assumption that *linear* relationships exist between flux and exciting forces of windings. This assumption allows us to use linear circuit analysis methods in model development. If this assumption is not valid, the values of the inductive elements within the model will be inaccurate and erroneous results will be obtained from its use. As an illustration of a violation of this assumption, consider the ϕ versus F plot shown in Fig. 12.6.

Here, we have depicted a situation where a magnetic material with the indicated characteristic is excited by two mmfs which are additive. It is seen that the resulting flux caused by the total mmf does *not* equal the sum of individual flux magnitudes. Therefore, to insure that the linear principle of superposition of flux values will always hold true, operation well below the saturation limits of the magnetic material is always necessary. One is therefore cautioned to always check that the point of maximum core excitation does not fall in a nonlinear portion of the material's characteristic before attempting to construct any linear electric circuit analog.

12.3. TWO-WINDING INDUCTORS AND THE ZERO-RIPPLE CURRENT PHENOMENON

During lab experiments at the California Institute of Technology in 1977, Dr. Slobodan Ćuk decided to wind both of the inductors of his SPC on a common magnetic core. The purpose of this particular experiment was to see if a savings in the weight and size of a basic Ćuk converter could be achieved without affecting its basic electrical properties. The experiment was successful, but an unexpected and potentially useful effect also occurred. While observing the input and output ripple current waveforms, Dr. Ćuk noted that they were very unequal, to the point where the magnitude in one was close to *zero* while the other remained near a value that would be expected if the two inductors were *not* coupled together on a common core. Also, the effect seemed to be *independent* of input and output voltage values, the switching frequency of the converter, switch duty cycle, and inductor current waveform (i.e., continuous or discontinuous)!

Since the time of these experiments, this peculiar effect that Dr. Ćuk observed has come to be known as the *zero-ripple current phenomenon*. Its appearance is dependent on the proper phasing of the voltages appearing across the terminals of the inductors as well as on certain details of the magnetic design of the dual-winding inductor. As it happens, the Ćuk converter inherently produces voltages across its inductors that are *always* proportional in magnitude and are *always* in phase. These waveform relationships are shown in Fig. 12.7, along with the schematic of a basic coupled-inductor Ćuk converter taken from Chapter 7 (Fig. 7.9B).

The voltage relationships of Fig. 12.7, along with the right choice of leakage inductances associated with the inductor windings, are the key factors in achieving zero-ripple current at *either* the input or the output of the Ćuk converter. Since a two-winding "coupled-inductor" is nothing more than a dual-winding transformer, the modeling example of Fig. 12.5D can be used to determine what values of leakage inductances are relevant to producing this unusual condition.

To make the following examination general, we will excite both the input and output of our transformer model with proportional voltage sources of like frequency and phase, as shown in Fig. 12.8. These conditions will be the only ones we will impose, and the waveform shapes of the voltage sources can take any form. For example, the shapes could be quasi-squarewave or sinusoidal as illustrated, differing only in amplitude by a factor defined here as a, where $a > 0$. For the Ćuk converter in Fig. 12.7, $a = 1$, as indicated by the accompanying voltage waveforms.

Let us now make the assumption that is possible to reduce the value of i_o to *zero* in our model of Fig. 12.8, and then examine the circuit currents and voltages that result from our assumption. If i_o is zero, then the AC voltage drop across L_2 must also be zero, as illustrated in Fig. 12.9A. The voltage appearing across L_C must therefore be equal to that of the secondary voltage source, av_s, reflected through the ideal transformer in the model. Also, the current through L_1 and L_C must be equal to that produced by the primary voltage source, v_s. These conditions lead to the simplified circuit model of Fig. 12.9B, where:

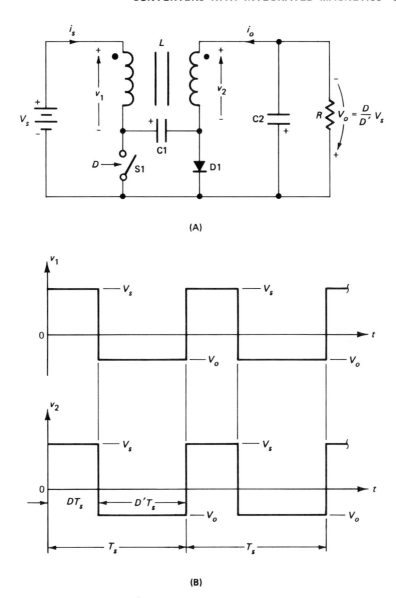

(A)

(B)

Fig. 12.7. A coupled-inductor Ćuk converter with equal primary and secondary voltage waveforms.

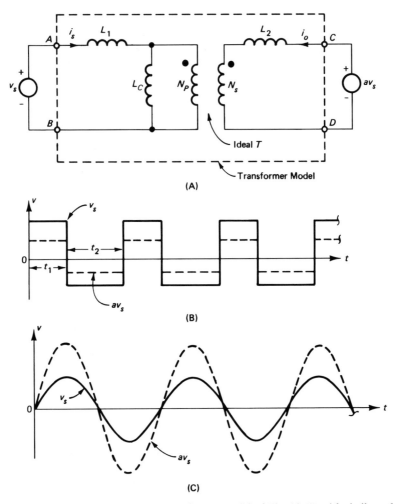

Fig. 12.8. Exciting the two-winding transformer model of Fig. 12.5D with similar voltage sources of equal frequency and phase.

$$v_s = (L_1 + L_C) \frac{di_s}{dt} \qquad (12.11)$$

$$\left[\frac{N_P}{N_s} \right] [av_s] = L_C \frac{di_s}{dt}. \qquad (12.12)$$

Combining Eqs. (12.11) and (12.12) by the elimination of the common di_s/dt factor gives:

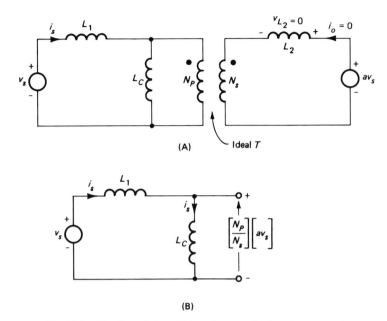

Fig. 12.9. Circuit model conditions for zero ripple output current.

$$L_1 = L_C \left[\frac{N_s}{aN_P} - 1 \right] . \tag{12.13}$$

The relationship between primary leakage inductance and that of the core material set by Eq. (12.13) is important and worth dwelling on for a moment. First of all, recall that this equation was derived based on the premise that *no* output ripple current existed in our model and, if satisfied, must produce this condition. It also indicates that a similar relationship can be established between L_2 and L_C to permit the input rather than the output ripple current to be reduced to zero. Using a similar path of analysis as depicted in Figs. 12.10A and B, this condition should occur when:

$$av_s = \left[L_2 + \left(\frac{N_s}{N_P} \right)^2 L_C \right] \frac{di_o}{dt} \tag{12.14}$$

$$\left[\frac{N_s}{N_P} \right] v_s = \left[\frac{N_s}{N_P} \right]^2 L_C \frac{di_o}{dt} \tag{12.15}$$

$$L_2 = \left(\frac{N_s}{N_P} \right)^2 L_C \left[\frac{aN_P}{N_s} - 1 \right] . \tag{12.16}$$

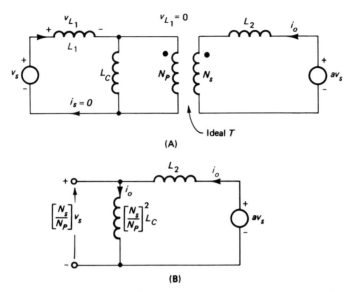

Fig. 12.10. Circuit model conditions for zero ripple input current.

Comparing the constraints imposed on the values of L_1 and L_2 by Eqs. (12.13) and (12.16), respectively, one finds that it is *not* possible to achieve zero input *and* output ripple currents simultaneously. For values of $N_s/aN_P > 1$, L_1 can be selected for zero output ripple. On the other hand, if $aN_P/N_s < 1$, then a value of L_2 can be found to reduce the input current ripple to zero.

Fig. 12.11 shows the current waveshapes that result when the conditions of either Eq. (12.13) or Eq. (12.16) are satisfied. Here, we have assumed that the input and output voltage sources take the form of quasi-squarewaves, such as those illustrated earlier in Fig. 12.8. Note from the equations accompanying Fig. 12.11 that the effective inductance seen at the primary side of the coupled-inductor is equal to its open-circuit value (i.e., secondary side open) when L_1 is selected for zero output current ripple. Conversely, the effective inductance seen at the secondary terminals will equal its open-circuit magnitude when L_2 is selected to achieve zero input current ripple.

In practice, it is difficult to design and consistently manufacture a transformer with specific values of leakage inductances to achieve the "matching" conditions specified by Eq. (12.13) or Eq. (12.14). One viable solution to this problem is to tightly wind primary and secondary turns together to reduce L_1 and L_2 to essentially zero, and then insert a small *external* "trimming" inductor in series with the input or output to emulate the desired value of L_1 or L_2.

This technique is shown in Fig. 12.12, where L_{ext} represents the external inductance needed for zero input (Fig. 12.12B) or output (Fig. 12.12A) current

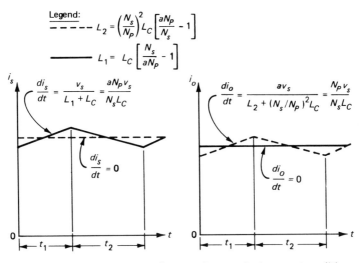

Fig. 12.11. Winding current waveshapes under zero ripple current conditions.

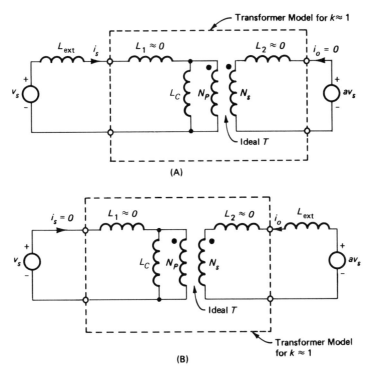

Fig. 12.12. Using external inductors to simulate leakage inductance required for zero current ripple.

ripple. Eqs. (12.13) or (12.16) can be used to calculate the ideal value of L_{ext} needed in either case. The actual contribution to this calculated value by the appropriate leakage inductance within the transformer can be found by empirical means. This entails measurement of the inductance of one winding while maintaining a low-inductance short across the other winding. Because the short will bypass the magnetization inductance of the transformer, only leakage inductance will be measured by this method.

12.4. APPLICATION IN BASIC CONVERTERS

At first thought, the zero-ripple current phenomenon may seem to be a scientific curiosity with little or no value in "real-world" SPC design. However, the ability to selectively reduce ripple currents of inductors within converters can be very advantageous when one considers that most SPC applications require low conducted noise on their input and output lines. Also, low ripple currents in SPC inductors reduce heating losses in associated converter capacitors, and often will decrease the values of capacitance needed for proper filtering of SPC input and/or output voltages.

From our generalized analysis of the last section, it is apparent that the zero-ripple current phenomenon can be produced in *any* converter whose inductances have proportional voltages of equal shape, frequency and phase, and is *not* just a special effect unique to the Ćuk converter circuit. To illustrate this point, let us pause and take a look at a few basic SPC design examples.

In Fig. 12.13, the output inductor of a basic buck converter has been replaced by one with two windings, N_s and N_P. Note that a second capacitor (C2) has been added to provide DC isolation for the "primary" winding of the inductor. Because of its position in the converter, this capacitance will see an average DC potential equal to that of the output voltage, V_o. Therefore, when S1 or D1 is conducting, both windings of L are forced to have *identical* voltage waveforms of *equal* magnitude. With a knowledge of the inductor's value (L_C), N_s, N_P, and L_{ext} can now be selected to reduce the dynamic "secondary" winding current to zero, using the criteria of Eq. (12.13) for $a = 1$.

The reader will recall that the converter of Fig. 12.13 was shown earlier in Chapter 8 (Fig. 8.5A) and can be implemented in a tapped form if desired. Although an output capacitor is shown (C1), it is not needed for ripple filtering. In practice, some capacitance is usually placed across the output of the converter to reduce second-order noise or to provide an instantaneous source of energy for dynamic loads.

Capacitor C2, on the other hand, is always required in this ripple reduction approach. Note that the elimination of the ripple current in N_s is produced by "steering" this dynamic current to the primary winding, N_P. Therefore, C2 can see significant dynamic current magnitudes and must be selected in size and value to properly handle them without excessive ESR power loss.

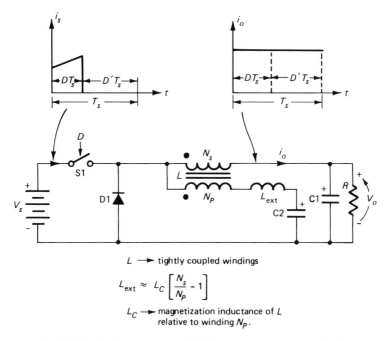

L ⟶ tightly coupled windings

$$L_{ext} \approx L_C \left[\frac{N_s}{N_p} - 1 \right]$$

L_C ⟶ magnetization inductance of L relative to winding N_p.

Fig. 12.13. Buck converter modified for zero output current ripple.

A basic boost converter can be altered in a similar fashion to achieve zero input ripple current. The resulting circuit is shown in Fig. 12.14. In this case, winding N_2 of the inductor is DC-isolated by capacitor C1, which sees a DC potential equal to that of the source voltage, V_s. Like the previous buck example, this isolation capacitor can see significant ripple current and its value and size must be chosen with care.

When two or more filter inductors of a converter see proportional equal voltages of equal frequency and phase, they can be placed on a common magnetic core and ripple currents adjusted so as to reduce one or more of them to essentially zero. Fig. 12.15 illustrates a coupled-inductor application in a two-output quasi-squarewave SPC. Here, the inductor of the second output, along with L_{ext}, is used to reduce the ripple current of the first output to zero. Again, we can use Eq. (12.13) to select L_{ext}; however, the proportionality constant, a, must now be set to approximately N_{s1}/N_{s2} of T1. The selection equation for L_{ext} accompanying Fig. 12.15 assumes that the four secondary diodes have negligible voltage drops and that T1 is an ideal transformer. Should either of these two assumptions not be valid, then the selection criteria for L_{ext} must be adjusted accordingly. The reader is referred back to Section 8.2 of Chapter 8 for further discussions of nonideal cross regulation of transformer windings and the effects of diode drops when dealing with SPCs with multiple unregulated outputs.

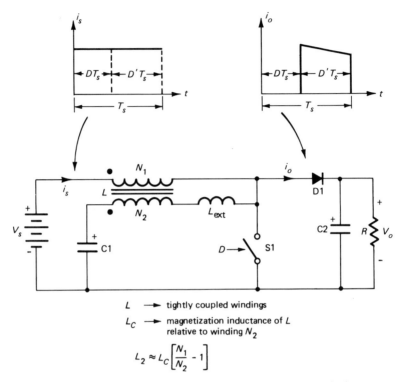

Fig. 12.14. Boost converter modified for zero input current ripple.

Fig. 12.16 shows one version of a basic Ćuk converter which uses two sets of coupled inductors to reduce *both* input and output ripple currents to zero. Here, both inductor assemblies require separate trimming inductances as well as decoupling capacitors. Another version of this same converter (Ref. 10) is shown in Fig. 12.17. In this case, by separating the energy transfer capacitance (C1 in Fig. 12.16) into two series parts, a common tie point for DC isolation of the secondary windings of both inductors is created. As a result, this zero-ripple Ćuk converter requires one less DC isolation capacitor than the circuit of Fig. 12.16.

12.5. THREE-WINDING INDUCTORS AND THE ZERO-RIPPLE PHENOMENON

Looking back at our examples of the last section, we see that a separate "tuning" winding was required to adjust the current ripple of a selected input or output inductor to zero. It is natural to inquire at this point if it is possible to devise a magnetic assembly whereby a *single* winding can *simultaneously* control the ripple current magnitudes in two other contemporary windings.

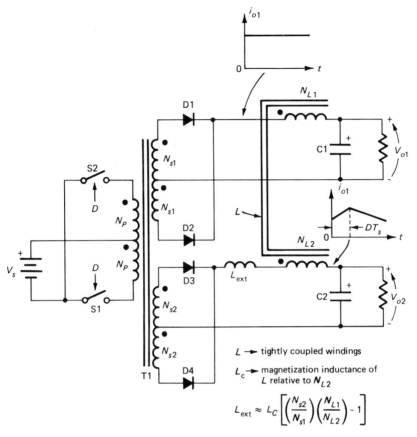

Fig. 12.15. Coupled-inductor quasi-squarewave converter with one output set for zero current ripple.

This combinational possibility is illustrated in a sequential fashion in Fig. 12.18. In part A, two separate magnetic assemblies are shown, each with two windings, with N selected as the ripple control winding in each case. A gap has also been placed into each core, as is the usual case for an inductor that will see a DC bias. In part B, we have placed the two cores from part A together with a common winding N. Finally, in part C, the two cores are combined into one structure, with winding N placed on its center portion.

Examination of the major flux paths illustrated in each part of Fig. 12.18 indicates that all three magnetic arrangements should be equivalent from an electrical standpoint, and that the single-core system of Fig. 12.18C is a viable scheme to permit winding N to control ripple current magnitudes in *both* winding N_1 and in winding N_2. To verify this idea, we will need to take a closer

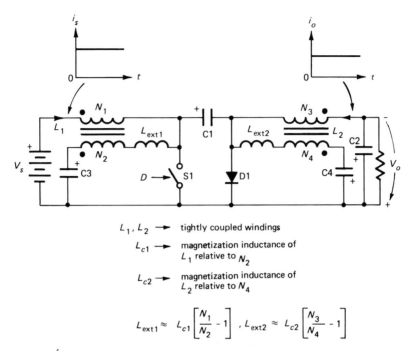

Fig. 12.16. Ćuk converter variation using two sets of coupled inductors to achieve zero input and output ripple.

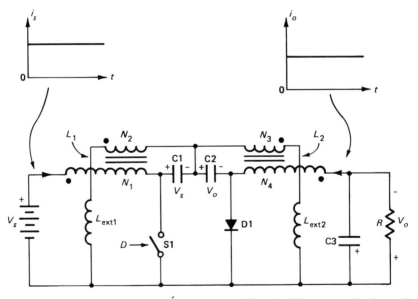

Fig. 12.17. Another variation of the Ćuk converter of Fig. 12.16 that uses one less decoupling capacitor.

284

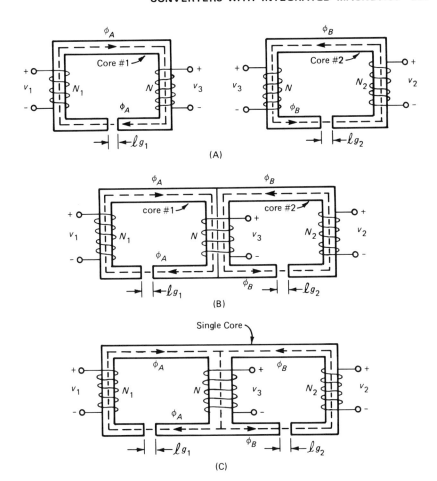

Fig. 12.18. Evolution of a single-core three-winding magnetic concept for zero current ripple control.

analytical look at the magnetic system of Fig. 12.18C in much the same manner that was followed for the two-winding inductor in Section 12.3.

We begin by considering a similar three-winding magnetic assembly illustrated in Fig. 12.19A. In this instance, an I core is used as an interface between two C cores, with air gaps introduced at each of the four joints. For our analysis, we will assume that all three cores have identical permeability and cross-sectional areas, and that the two sets of air gap lengths are uniform with no fringing fluxes associated with them. Two major and three leakage flux paths are also assumed, with directions as shown, and each of the three windings is driven by a zero-impedance voltage source.

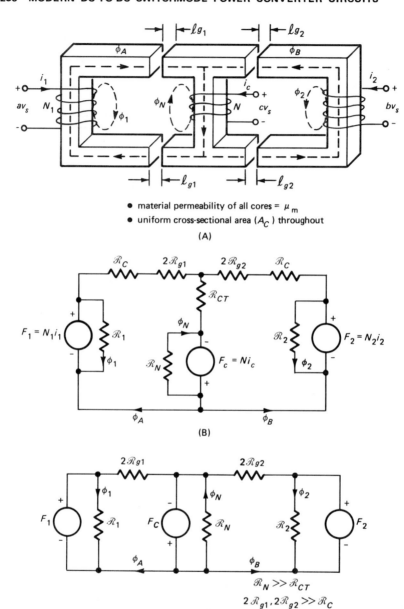

- material permeability of all cores = μ_m
- uniform cross-sectional area (A_C) throughout

(A)

(B)

$$\mathscr{R}_N \gg \mathscr{R}_{CT}$$
$$2\mathscr{R}_{g1}, 2\mathscr{R}_{g2} \gg \mathscr{R}_C$$

(C)

Fig. 12.19. Developing a reluctance and mmf model for the magnetic of Fig. 12.18C.

Again, we assume that all three voltage sources (av_s, bv_s, cv_s) are of similar shape, proportional in amplitude (set by constants a, b, and c), have the same frequency and phase, and constants $a, b, c > 0$.

With all important electrical and magnetic circuit conditions established, we now proceed as before to establish an equivalent circuit model. First, the basic reluctance and mmf topology is drawn with flux paths taken from Fig. 12.19A. This step is shown in Fig. 12.19B, with all series reluctances lumped together. $\Re_C$ represents the total reluctance of each of two C core paths, with $\Re_1$, $\Re_2$, and $\Re_N$ as those to related winding leakage paths. $\Re_{CT}$ is the reluctance of the I core material, with air gap reluctances designated by $\Re_{g1}$ and $\Re_{g2}$. For simplicity, we have omitted the actual equations here (as well as in future model developments of this chapter) for relating reluctance values to the mean dimensions of the cores and associated air gap lengths. For those interested in these details, they can be easily calculated using the rules followed earlier in the development of a circuit model for a simple inductor (see Figs. 12.2B and C).

To further simplify our work, we can also reasonably assume that the air gap reluctances will be much greater than those presented by the core materials. The elimination of $\Re_C$ and $\Re_{CT}$ from Fig. 12.19B results in the reduced reluctance model of Fig. 12.18C. Taking the dual of this simplified model produces the permeance network of Fig. 12.20A. Using the av_s winding as a reference, elements of this network are then scaled by N_1, resulting in a circuit model relating flux linkages to winding currents (Fig. 12.20B). Finally, all permeances are replaced by their corresponding inductance values and all flux linkages by voltage sources, scaled as necessary using ideal transformers with proper turns ratios (Fig. 12.20C).

We can further manipulate the electrical network model of Fig. 12.20C to place the leakage inductances associated with windings N and N_2 on the "primary" side of their respective ideal transformer. Performing this impedance translation results in the final network model shown in Fig. 12.21.

Once again, using the same simple procedure for model development, we have been able to reduce a rather complex magnetic circuit to a simplified electrically equivalent network. This network can now be used to establish the electrical conditions necessary to produce zero-ripple currents in the two outer windings, N_1 and N_2.

First, we see from the model of Fig. 12.21 that, if currents i_1 and i_2 have been reduced to zero, there can be *no* dynamic voltage drops across leakage inductances L_1 and L_2. This observation also implies that L_1 and L_2 could assume *any* values and still not be important in the determination of what model values are needed to produce zero ripple current conditions in their respective windings. It also can be deduced the values of the remaining model circuit elements, namely air gap inductances and the leakage inductance of the center winding, *will* be important in the production of the zero-ripple phenomenon.

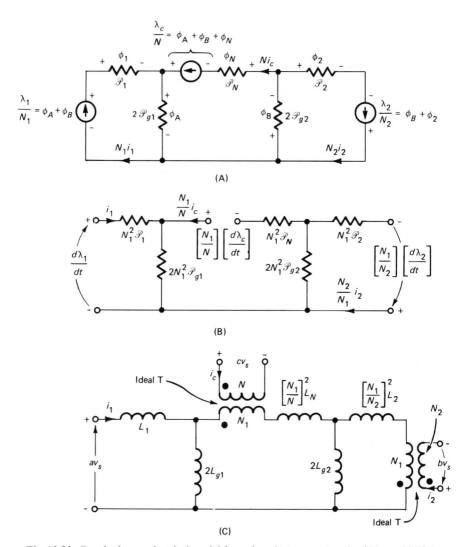

Fig. 12.20. Developing an electrical model from the reluctance network of Figure 12.19C.

With i_1 and i_2 equal to zero, we can redraw the model network of Fig. 12.21 to depict the circuit conditions that result from their absence. When this is done, the simplified model of Fig. 12.22 is obtained. Note that, with these currents set at zero, the voltages appearing across the gap inductances are, at most, simple reflections of associated winding voltages since their leakage inductance voltage drops must also be zero. We can now establish the relationships between L_N and that of the two gap inductances by simple network analysis as follows:

$$av_s = \left(\frac{N}{N_1}\right)(2L_{g1})\frac{di_c}{dt} \tag{12.17}$$

$$b\left[\frac{N_1}{N_2}\right]v_s = \left(\frac{N}{N_1}\right)(2L_{g2})\frac{di_c}{dt} \tag{12.18}$$

$$cv_s - v_x = L_N\frac{di_c}{dt} \tag{12.19}$$

$$v_x = \frac{N}{N_1}\left[av_s + \left(\frac{N_1}{N_2}\right)bv_s\right]. \tag{12.20}$$

Combining Eqs. (12.17) and (12.18) with Eqs. (12.19) and (12.20) and eliminating the common di_c/dt terms gives:

$$\frac{L_N}{2L_{g1}} = \left(\frac{c}{a}\right)\left(\frac{N_1}{N}\right) - \left(\frac{b}{a}\right)\left(\frac{N_1}{N_2}\right) - 1 \tag{12.21}$$

$$\frac{L_N}{2L_{g2}} = \left(\frac{c}{b}\right)\left(\frac{N_2}{N}\right) - \left(\frac{a}{b}\right)\left(\frac{N_2}{N_1}\right) - 1. \tag{12.22}$$

Therefore, if the conditions set by Eqs. (12.21) and (12.22) are *simultaneously* satisfied, then the ripple currents in *both* windings N_1 and N_2 will be reduced to

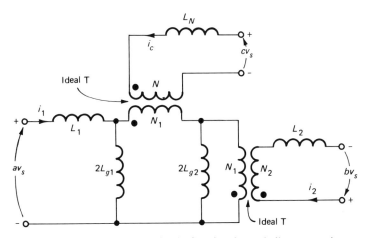

Fig. 12.21. Final equivalent electrical circuit for the three-winding magnetic system of Fig. 12.19A.

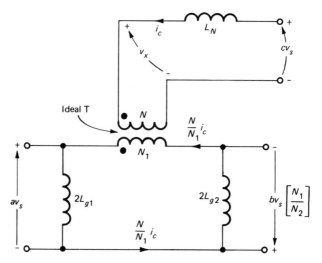

Fig. 12.22. Circuit conditions within the model of Fig. 12.21 when both i_1 and i_2 equal zero.

zero. Also, the effective inductance seen at the terminals of the center winding N will be equal to

$$L_e = L_N + \left(\frac{N}{N_1}\right)^2 [2L_{g1} + 2L_{g2}] \tag{12.23}$$

under these same conditions. Unlike the zero-ripple dual-winding inductor design criteria given in Section 12.3, note that Eqs. (12.21) and (12.22) are *not* interdependent. Therefore, assuming that winding turns and proportionality constants are known, values for the gap inductances L_{g1} and L_{g2} can be found that satisfy both of these equations for any fixed value of center leg leakage inductance, L_N.

For example, assume that we have a converter application where $a = 1$, $b = 2$, $c = 3$, and $N_1 = N_2$. Placing these values into Eqs. (12.21) and (12.22), one finds that gap inductance L_{g2} must be twice L_{g1}, and that the choice of N, N_1, and L_N will determine the actual gap inductance values needed to produce zero ripple currents in both N_1 and N_2.

Referring once again to the network of Fig. 12.22, we see that we can increase the effective value of the leakage inductance L_N by simply adding a small amount of *external* inductance in series with it, just as we did in the case of the two-winding inductor in Section 12.3. In actual design practice, it may be difficult to predict and maintain a constant value for L_N since it is essentially a parasitic component; therefore, supplementing it by external means could be very desirable, if not mandatory, for a production application.

With the conditions established to achieve current ripple reduction in two of the three windings of the assembly shown in Fig. 12.19A, let's take a look now at two applications of this unique magnetic in typical converter circuits.

12.6. MORE ZERO-RIPPLE CONVERTER APPLICATIONS

Figure 12.23 shows a three-output buck-derived quasi-squarewave converter with all output inductances wound on a common core. In this case, the ripple currents in the first and third outputs are controlled by the common inductance associated with the second output. Since the voltage waveforms across each of the three inductances are of the same shape, phase, and frequency, we can apply Eqs. (12.21) and (12.22) to select the air gap lengths and the leakage inductance of the center leg of the core to achieve zero ripple in the outer windings. In this application, the constants a, b, and c are established by the corresponding primary-

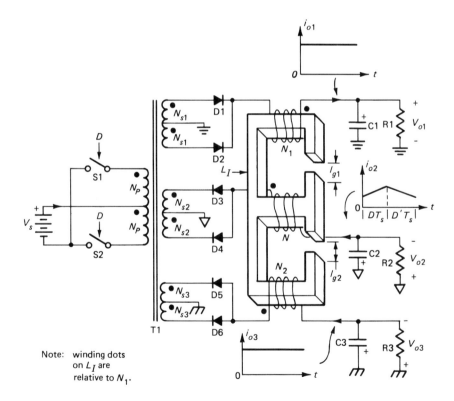

Fig. 12.23. A coupled-inductor three-output quasi-squarewave SPC with two outputs adjusted for zero current ripple.

to-secondary turns ratio of T1 as:

$$a = \frac{N_{s1}}{N_P} \qquad (12.24)$$

$$b = \frac{N_{s3}}{N_P} \qquad (12.25)$$

$$c = \frac{N_{s2}}{N_P}. \qquad (12.26)$$

Placing the above values for a, b, and c into Eqs. (12.21) and (12.22), and noting that we have only one air gap on each side of the center leg of the core in Fig. 12.23, gives:

$$\frac{L_N}{L_{g1}} = \left(\frac{N_{s2}}{N_{s1}}\right)\left(\frac{N_1}{N}\right) - \left(\frac{N_{s3}}{N_{s1}}\right)\left(\frac{N_1}{N_2}\right) - 1 \qquad (12.27)$$

$$\frac{L_N}{L_{g2}} = \left(\frac{N_{s2}}{N_{s3}}\right)\left(\frac{N_2}{N}\right) - \left(\frac{N_{s1}}{N_{s3}}\right)\left(\frac{N_2}{N_1}\right) - 1. \qquad (12.28)$$

The above equations assume that the various secondary diode drops are negligible and that the converter's transformer (T1) is an ideal circuit element. Therefore, it may be necessary to alter the values for a, b, and c to account for any deviation from this assumption, should these diode voltages or transformer parasitic contributions be significant. The reader is referred once again to Chapter 8 for more details of this particular problem.

Of all the various converter circuits, perhaps the Cuk converter benefits the most from the use of an integrated magnetic design such as that shown earlier in Fig. 12.19. One such circuit variation (Ref. 11) is illustrated in Fig. 12.24. Here, the main energy transfer capacitance of the converter has been split into two series parts. Note that *two* isolated windings of equal turns (N) have been placed on the center leg of the magnetic assembly, with one winding tied to C1 and the other to C2. With these placements, both of the center-leg windings will *always* see a voltage waveform *identical* to that appearing across both outer windings (N_1 and N_2) of the magnetic assembly. Therefore, these two inner windings form a one-to-one transformer element, providing a means of DC isolation between V_s and V_o as well as ripple current control of both input and output lines.

The equations accompanying Fig. 12.24 give the criteria for selecting the two air gap lengths and are based on the conditions required by Eqs. (12.21) and (12.22). In many practical applications of this particular converter, the outer windings usually have the same number of turns ($N_1 = N_2$).

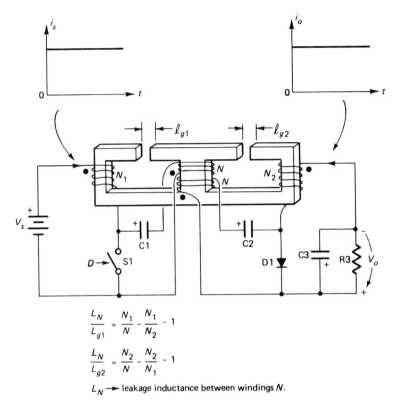

$$\frac{L_N}{L_{g1}} = \frac{N_1}{N} \cdot \frac{N_1}{N_2} - 1$$

$$\frac{L_N}{L_{g2}} = \frac{N_2}{N} \cdot \frac{N_2}{N_1} - 1$$

$L_N \longrightarrow$ leakage inductance between windings N.

Fig. 12.24. Ćuk converter design that emulates a true DC transformer with turns ratio set by duty cycle D.

For this special case, our selection equations reduce to:

$$\frac{L_N}{L_{g1}} = \frac{N_1}{N} - 2 \qquad\qquad (12.29)$$

$$\frac{L_N}{L_{g2}} = \frac{N_1}{N} - 2 \qquad\qquad (12.30)$$

since $a = b = c = 1$. Therefore, it is apparent from Eqs. (12.29) and (12.30) that the two core air gap lengths will be *equal* for this particular case, again assuming that core cross-sectional areas are uniform and equal throughout.

Let us now pause for a moment and reflect on the significance of the Ćuk converter design of Fig. 12.24. Here, an SPC variation has been created which emulates a *true DC transformer*. The "turns ratio" of this DC transformer is set by the value of the switch duty cycle and is therefore *electronically variable*.

Both input and output currents can be made to be *nonpulsating* and will remain that way regardless of changes in converter switching frequency, switch or diode duty cycle, and load. In addition, all of its magnetics are contained within a *single* magnetic assembly, including a transformer for DC isolation of input and output voltages!

12.7. A FORWARD CONVERTER WITH INTEGRATED MAGNETICS

Up to this point, we have restricted our discussions of magnetic integration methods to those that directly relate to inductances with proportionally equal voltages impressed across them. With the single exception of the Ćuk converter variation just discussed (Fig. 12.24), we have excluded AC transformers from our design development of integrated magnetic SPCs. In the special case of this Ćuk-derived circuit, it was possible to fashion a single magnetic assembly which housed not only the two inductors of the converter, but also a 1:1 isolation transformer. All this was done with no compromise in conversion characteristics or performance.

The Ćuk converter is a unique SPC topology in that all inductor as well as any transformer windings must *always* have proportional voltages of equal shape, frequency, and phase. However, other converters are not as fortunate in this respect, and it is not readily apparent how one can magnetically combine inductive as well as AC transformer components in these circuits.

For example, consider the case of a buck-derived forward converter using the circuit variation shown in Fig. 5.20B of Chapter 5. Here, the windings of the isolation transformer T1 see voltages proportional to the source potential, V_s, while its inductor L sees voltages proportional to the difference between V_s and the output voltage, V_o. Since V_o is a function of switch duty cycle (D), the inductor's AC voltage will fluctuate in accord with changes in D. Similar observations can be made of transformer and inductor voltage waveforms in boost-derived converters, such as the single-ended design shown in Fig. 6.14 of Chapter 6.

However, we do know that the AC transformers and inductors of converters share a common dependency on flux change in ferromagnetic materials to perform their functions. Recall from the basic definition of inductance set by Eqs. (12.1) and (12.2) that flux change in any magnetic medium can be related to current variation as:

$$v = L \frac{di}{dt} = N \frac{d\phi}{dt} \qquad (12.31)$$

or

$$d\phi = \frac{L}{N} di. \qquad (12.32)$$

We also remember from our previous modeling experiences of this chapter that many magnetic core structures have multiple flux paths. Depending on winding arrangements, it is therefore feasible to devise core structures wherein various components of flux add or substract from one another in some portions of its magnetic paths. Therefore, it would seem very probable that an integrated magnetic assembly could be conceived that would emulate as well as properly relate various flux changes one would expect in the inductors and transformers of an SPC.

Since it is possible by network analysis to formulate a set of equations that relate the various dynamic currents to circuit voltages for each conduction state of a converter, we can then modify them to relate currents to magnetic flux changes, using the conditions of Eq. (12.32). The modified equations then show how the flux changes of various magnetics of an SPC are interrelated and act as guides in constructing an integrated magnetic core system that satisfies the re-quired relationships.

As an illustration of an SPC that was conceived using the above developmental method, let us examine the workings of a single-magnetic converter shown in Fig. 12.25. Here, two E cores have been joined with an air gap introduced only in the center leg section of the converter's magnetic component. At first glance, this SPC topology shows some resemblance to that of a forward converter; how-ever, the output inductance used for energy storage when the switch (Q1) is in an OFF state is not readily discernable.

We can use the modeling tools of Section 12.1 to analyze the converter of

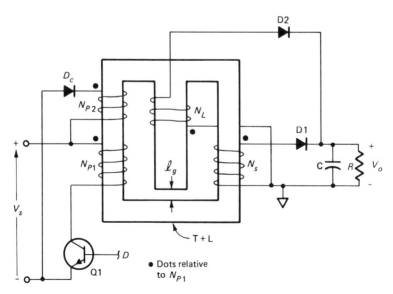

Fig. 12.25. An integrated magnetic forward converter.

Fig. 12.25 to see if this inductor is "hidden" within its magnetic assembly. First, we isolate the magnetic and identify winding potentials and current directions. This is done in Fig. 12.26, with current directions corresponding to those that would result if the SPC switch and/or diodes were in a conducting state. Major flux paths and their directions resulting from the indicated winding currents are also identified, using the right-hand rule and the procedure given in Section 12.1 for "dotting" of the five windings relative to N_{P1}. Because we are interested only in the major flux paths for this investigation, secondary leakage flux paths are not considered here; however, they can be added at the expense of a more complicated model, just as was done for the examples in Figs. 12.4 and 12.19.

Next, the reluctance model for the magnetic assembly is drawn in Fig. 12.27A using the information contained in Fig. 12.26. Since the core structure is symmetrical and the air gap reluctance will be dominant in the center leg, the circuit model of Fig. 12.27A can then be simplified to that shown in Fig. 12.27B. Note also that the two left-hand mmfs can be combined into a single mmf because of their series connection in one leg of the core structure.

Once again, we can use duality relationships to convert the network of Fig.

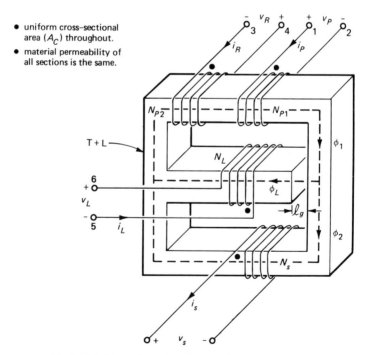

Fig. 12.26. Magnetic used in the SPC design of Fig. 12.25.

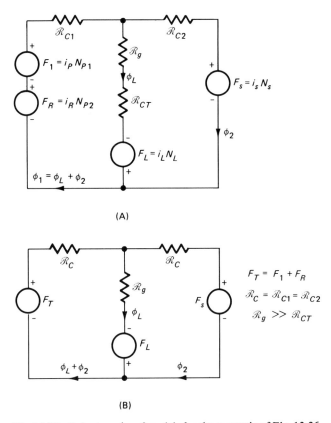

(A)

(B)

Fig. 12.27. Reluctance/mmf models for the magnetic of Fig. 12.26.

12.27B to one involving permeances, as is done in Fig. 12.28A. This new network is then scaled to the winding chosen as a reference. The circuit of Fig. 12.28B results from this scaling step, with N_{P1} designated as the reference winding. Finally, the scaled permeances are then replaced by inductances and ideal transformers added as required to refer winding currents to their terminal values, as shown in Fig. 12.28C.

We now have an electrical model for the magnetic system of Fig. 12.26. Substituting this model into the converter of Fig. 12.25 results in the equivalent SPC circuit shown in Fig. 12.29. Note that, in making this substitution, we have also moved the gap inductance to the secondary side of the ideal transformer of turns ratio, $N_{P1} : N_s$. In making this impedance translation, the ideal transformer of turns ratio $(N_{P1} : N_L)$ in Fig. 12.28C must also be moved and its ratio changed as indicated in Fig. 12.29.

The converter of Fig. 12.29 is now beginning to resemble a forward converter.

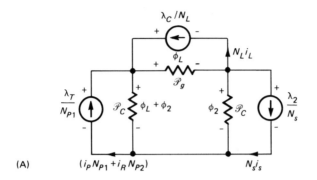

(A)

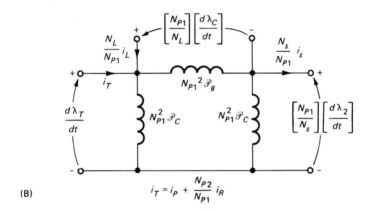

(B)

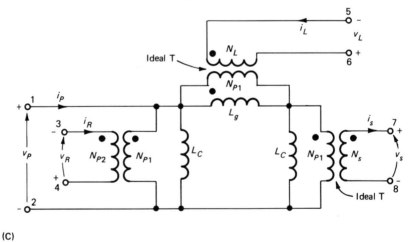

(C)

Fig. 12.28. Development of an electrical circuit equivalent for the magnetic of Fig. 12.26 using the reluctance network of Fig. 12.27B.

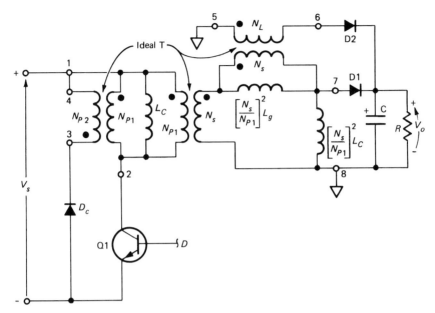

Fig. 12.29. Complete electrical circuit for the SPC of Fig. 12.25.

If we now assume that L_C is much greater than L_g and that $N_L = N_s$, the converter circuit can be further simplified to that of Fig. 12.30A. Note that we have also converted the ideal transformer of turns ratio $(N_{P1}:N_s)$ in Fig. 12.29 to an imperfect one that has a finite value of magnetizing inductance set equal to L_c.

The topology of Fig. 12.30A can be reduced further by the elimination of the two remaining ideal transformers. The clamp winding on the primary side of the converter can be placed within T1 if N_{P1} and N_{P2} are wound tightly coupled. On the secondary side of the converter, the ideal 1 : 1 transformer associated with the reflected center leg inductance can be removed by manipulating the circuit positions of diodes D1 and D2 without compromising their conduction intervals.

The final converter structure that results from the elimination of all ideal transformers from the circuit of Fig. 12.30A is illustrated in Fig. 12.30B. Comparing this latter equivalent SPC to the forward design of Fig. 5.20B in Chapter 5, we find that the two topologies are identical!

Therefore, we can conclude that the SPC shown earlier in Fig. 12.25 is indeed an integrated-magnetic version of a buck-derived forward converter. It is also apparent from our analysis that the air-gapped center leg of this unusual magnetic assembly serves as an energy storage medium to supply output power when the primary switch (Q1) is in an OFF state. When Q1 is ON, energy in the center leg is replenished as well as transferred from the input source to the output via the outer legs of the core structure. Winding N_{P2} is used to return any energy stored in the outer legs to the source when Q1 is turned OFF.

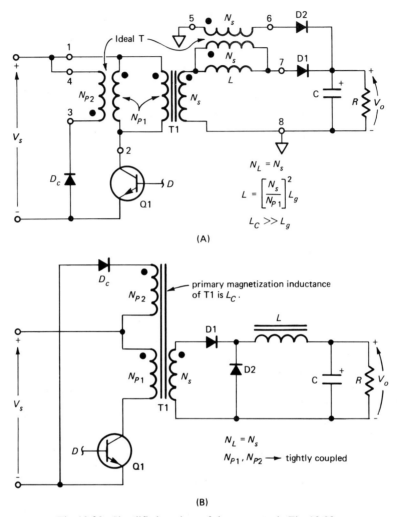

Fig. 12.30. Simplified versions of the converter in Fig. 12.29.

Assuming that the output current maintains a form that is continuous, the ideal voltage transfer function of this converter can be found by equating the volt-seconds experienced by the center leg winding during times DT and $D'T$. Ignoring all switch and diode voltage drops and assuming that conversion efficiency is 100%, the output-to-input voltage ratio is therefore found to be:

$$\frac{V_o}{V_s} = \frac{N_s}{N_{P1}} D, \quad N_L = N_s. \tag{12.33}$$

Note that the ideal voltage transfer function of Eq. (12.33) is *identical* to that determined earlier in this textbook for a forward converter with discrete transformer and inductor magnetic elements.

Examination of the current waveforms in the switch and the diodes of this converter also confirms that is a buck-derived SPC. A typical set of these waveforms are shown in Fig. 12.31A. Note that the magnetization energy of the outer legs is returned to the source during $D'T$ intervals via diode D_c. Although the total output current (i_o) is shown to be continuous at the converter switching transitions, some discontinuity may be seen because of the loss of this magnetization energy. This discontinuity can be minimized by making the ratio of outer leg inductance to center leg inductance as large as possible (say 10 times) so that magnetization energy level will be small in comparison to that delivered to the load of the converter.

Contemporary voltage waveforms that would be observed in this converter are shown in Fig. 12.31B. Here, again, they are indicative of the dynamic potentials of a buck-derived forward design operating in a continuous mode. The waveform shown for the center leg winding assumes a measurement across N_L

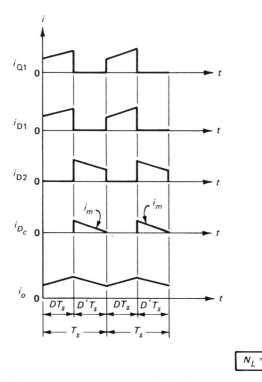

Fig. 12.31. (A) Typical current waveforms within the converter circuit of Fig. 12.25.

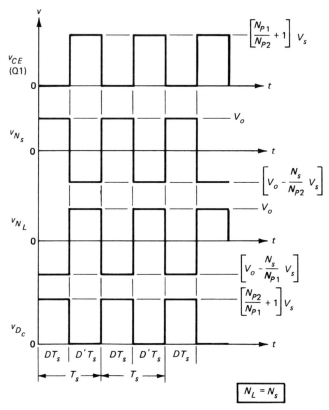

Fig. 12.31. (*Continued*) (b) Ideal voltage waveforms within the converter circuit of Fig. 12.25.

with ground reference at its "undotted" terminal, while that for diode D_c is taken at its cathode connection (terminal 3 in Fig. 12.29) with respect to the source voltage return.

Additional outputs can be accommodated by simply adding more inductor windings and commutating diodes on the center leg of the magnetic in Fig. 12.25, along with corresponding secondary windings and diodes placed on the outer leg. Also, in those applications where N_{P1} can be made equal to N_{P2}, then a single-primary version of the converter of Fig. 12.25 can be made. This SPC variation is shown in Fig. 12.32, where both switches (Q1 and Q2) are turned ON and OFF simultaneously. Diodes D3 and D4 then provide the path for return of magnetization energy to the source when both switches are OFF. The reader might recall that this particular core reset scheme was discussed previously in Chapter 4 (Fig. 4.10A). In fact, all of the core reset methods discussed in this earlier chapter are applicable to single-ended integrated magnetic SPC designs.

If we choose *not* to make N_L equal to N_s in the converter of Fig. 12.25, then the output current ripple will become somewhat discontinuous, just as might be expected in a forward converter with "tapped" inductor or transformer windings. The subject of magnetic tapping was covered in Chapter 8 and all alternatives discussed there for buck-derived SPCs apply as well to the integrated designs of Figs. 12.25 and 12.32.

12.8. AN ISOLATED BOOST SPC WITH INTEGRATED MAGNETICS

If we now apply the rules of bilateral inversion that we learned in Chapter 9 to the integrated forward design of Fig. 12.25, a dual boost converter approach can be derived. Fig. 12.33 shows the converter circuit that results from this duality exercise.

This particular SPC is an integrated magnetic version of a boost design shown earlier in Fig. 6.14A of Chapter 6. When switch Q2 is ON and switch Q1 is OFF, energy is stored in the air gap inductance of the center leg of the magnetic structure in Fig. 12.33. During these time intervals, the energy in the output capacitor C supplies load power needs. Also, any inductive energy stored in the outer legs of the structure is routed to the load via diode D3. When Q1 and Q2 reverse conduction states, the energy stored in the center leg is released to the output winding N through diode D1 and diode D3 turns OFF. With N_L set equal to N_P, the ideal voltage transfer function (continuous mode) for this converter is;

$$\frac{V_o}{V_s} = \frac{N_s}{N_P}\left[\frac{1}{1-D}\right] \tag{12.34}$$

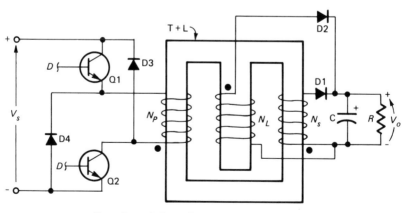

Note: Dots relative to N_P.

Fig. 12.32. One variation of the integrated magnetic forward converter shown in Fig. 12.25.

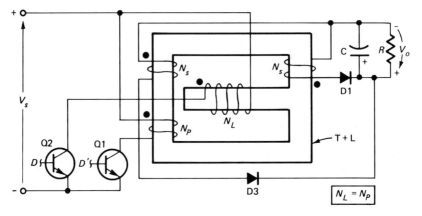

Note: Dots relative to N_p.

Fig. 12.33. An integrated magnetic boost converter.

which proves that this SPC is indeed a boost-derived design. This is to be expected, however, as the rules of bilateral inversion and duality must apply to all converter circuits, *including* those that have integrated magnetics.

12.9. RUDIMENTARY SYNTHESIS METHODS

Having established by analysis methods that the SPC circuits of Figs. 12.25 and 12.33 are integrated magnetic versions of basic buck and boost topologies with transformer isolation, let us now see if we can formulate an easy way to synthesize an integrated magnetic version of each of these SPCs, given the discrete-magnetic circuit arrangements as starting points.

To begin, we must first reconstruct the circuit schematics of these SPCs so as to detail the magnetic aspects of the transformer and inductor components. The schematics that result from this reconstruction process are illustrated in Figs. 12.34 and 12.35 for buck and boost versions, respectively. Note that a flux direction within each magnetic component has also been assigned, based on winding polarities produced by converter operation.

Next, for each of the two switching intervals of the converters (continuous mode of operation assumed), a set of equations defining the rate of change of flux in each magnetic component is established. These equations, of course, will be dependent on circuit voltage and the number of turns of related transformer and/or inductor windings. For this exercise, we can assume all semiconductors of the SPCs are ideal in order to simplify our work. Also, we will ignore potential leakage inductance effects between transformer windings for the same reason, and assume that fluxes are completely contained within its core structure.

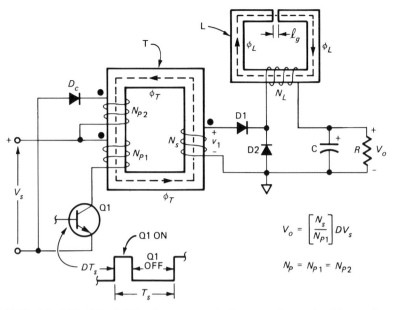

Fig. 12.34. The SPC of Fig. 5.20B redrawn to emphasize magnetic aspects of its transformer and inductor.

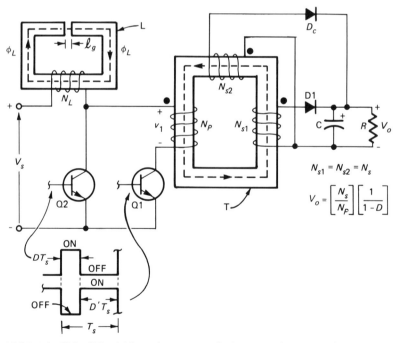

Fig. 12.35. The SPC of Fig. 6.14A redrawn to emphasize magnetic aspects of its transformer and inductor.

Thus, for the buck converter of Fig. 12.34 during interval DT:

$$\frac{d\phi_T}{dt} = \dot\phi_T = \frac{V_s}{N_P} = \frac{V_1}{N_s} \tag{12.35}$$

$$\frac{d\phi_L}{dt} = \dot\phi_L = \frac{V_1}{N_L} - \frac{V_o}{N_L} \tag{12.36}$$

where $\dot\phi_T$ is the rate of flux change in the transformer core, $\dot\phi_L$ is the rate of change of inductor flux, and V_1 is the peak value of secondary voltage during DT. For the interval $D'T$:

$$\frac{d\phi_T}{dt} = \dot\phi_T = \frac{V_s}{N_P} \tag{12.37}$$

$$\frac{d\phi_L}{dt} = \dot\phi_L = \frac{V_o}{N_L}. \tag{12.38}$$

Referring back to Eqs. (12.35) and (12.36), we can combine them to eliminate the dependent variable, V_1. Performing this combination gives:

$$\dot\phi_T = \left[\frac{N_L}{N_s}\right]\dot\phi_L + \frac{V_o}{N_s}. \tag{12.39}$$

Note that the last term of Eq. (12.39) is of a form that could be considered as defining a flux change in a magnetic medium that is dependent on the output voltage, V_o, of the converter and the number of secondary turns on the converter's transformer, N_s. Also, it follows that, if we make this consideration, that the magnetic medium *must* be a part of the transformer assembly itself in order that the constraints on flux changes defined by Eq. (12.29) are satisfied. Therefore, we can rewrite Eq. (12.39) as:

$$\dot\phi_T = \left[\frac{N_L}{N_s}\right]\dot\phi_L + \dot\phi_o \tag{12.40}$$

where $\dot\phi_o = V_o/N_s$. Turning now to the first term of Eq. (12.39), we note that its contribution to $\dot\phi_T$ is dependent on a fraction, N_L/N_s, of the rate of change of flux in the SPC's inductor. Since our goal is to make the inductor a part of the same magnetic assembly that contains the transformer component, it is logical to assume that N_L should be made equal to N_s, so that *all* of this flux change is contained within one magnetic path or leg of this assembly. The reader may also recall that this same assumption permitted us to reduce the rather complex magnetic model of Fig. 12.29 to a form that facilitated recognition of this SPC as a buck-derived converter. Therefore, setting $N_L = N_s$ in Eq. (12.40), we arrive at an expression for $\dot{\phi}_T$ as:

$$\dot{\phi}_T = \dot{\phi}_L + \dot{\phi}_o. \tag{12.41}$$

Remembering our magnetic modeling experiences from earlier sections of this chapter, we can interpret Eq. (12.41) as defining a magnetic assembly in which there are three major flux paths. It also tells us that the flux change in an input source-related path (ϕ_T) contributes to the change in another path associated with the inductor portion of the magnetic assembly (ϕ_L), as well as to flux change in a third path (ϕ_o). From Eq. (12.40), this third leg must be related to output voltage value.

These general observations permit us now to sketch out a magnetic path arrangement that satisfies the needs of Eq. (12.41). This is done in Fig. 12.36A. Note that we have added a gap in the "inductor" path area because we expect this leg will have significant DC bias, just as a discrete inductor in a buck converter would experience.

Having found a magnetic arrangement that meets the flux change relationships for the interval DT, we now use Eqs. (12.38) and (12.39) to establish similar relationships for the interval $D'T$. This is done keeping in mind the core arrangement of Fig. 12.36A is still needed for the interval DT. The resulting magnetic system is shown in Fig. 12.36B.

To complete our synthesis effort, we now combine the two magnetic arrangements of Fig. 12.36 into a single magnetic element, adding surrounding switches

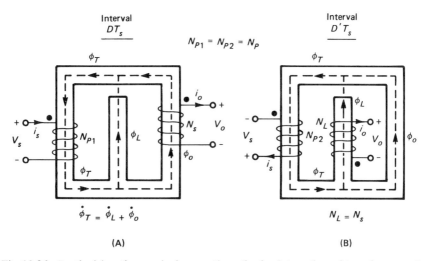

Fig. 12.36. Synthesizing the required magnetic paths for integration of transformer and inductor in a forward converter.

and diodes to satisfy the needs of each of the two switching states of the SPC. This final step results in the integrated magnetic forward converter shown earlier in Fig. 12.25 of this chapter!

The total synthesis procedure can be repeated for the boost converter of Fig. 6.14A from Chapter 6. First, the converter is redrawn as shown in Fig. 12.35, emphasizing the magnetic aspects of its inductor and its transformer. Second, the idealized flux change rates for each magnetic are found using this schematic. For the interval DT:

$$\frac{d\phi_L}{dt} = \dot{\phi}_L = \frac{V_s}{N_L} \tag{12.42}$$

$$\frac{d\phi_T}{dt} = \dot{\phi}_T = \frac{V_o}{N_s}. \tag{12.43}$$

Repeating for the interval $D'T$:

$$\frac{d\phi_L}{dt} = \dot{\phi}_L = \frac{V_1}{N_L} - \frac{V_s}{N_L} \tag{12.44}$$

$$\frac{d\phi_T}{dt} = \dot{\phi}_T = \frac{V_1}{N_P} = \frac{V_o}{N_s}. \tag{12.45}$$

We now combine Eqs. (12.44) and (12.45) to eliminate the intermediate variable, V_1, and set $\dot{\phi}_s = V_s/N_P$:

$$\dot{\phi}_T = \left[\frac{N_L}{N_P}\right] \dot{\phi}_L + \dot{\phi}_s. \tag{12.46}$$

As was the case for the forward design just discussed, Eq. (12.46) indicates that N_L should be made equal to N_P to contain all inductor flux change within one path of the magnetic assembly under consideration. Therefore:

$$\dot{\phi}_T = \dot{\phi}_L + \dot{\phi}_s, \qquad N_L = N_P. \tag{12.47}$$

Next, we use Eq. (12.47) as a guide in sketching a magnetic core arrangement for the interval $D'T$, adding a gap in the "inductor" leg for DC bias toleration. Another arrangement sketch is prepared for the interval DT, using the conditions of Eqs. (12.42) and (12.43) as guides. Finally, the two sketches are combined, along with SPC switches and diodes to set the magnetic states for each portion of the switching interval. The final circuit that results is (as we might expect at this point) identical to that shown in Fig. 12.33! The reader may recall that this particular integrated magnetic SPC was originally evolved by applying bilateral inversion to the SPC of Fig. 12.25.

This rudimentary synthesis procedure is fairly simple and straightforward to apply to any SPC circuit. While neither rigorous nor exact in results, it does yield an integrated magnetic version of any SPC quickly, and gives good insight as to the major magnetic aspects of the required core arrangements and construction needs.

12.10. MORE COMPLEX CONVERTERS WITH INTEGRATED MAGNETICS

The single-ended integrated magnetic forward converter of Fig. 12.25 can be modified so as to use two primary switch elements operating in a so-called "push-pull" manner. Two versions with this modification are shown in Fig. 12.37 (Ref. 12). In Fig. 12.37A, an additional set of primary and secondary windings have been added with "dot" polarities referenced to the primary winding controlled by switch Q1. In this version, when either of the primary switches is ON, energy is stored in the air gap of the magnetic's center leg. When Q1 is ON and Q2 is OFF, energy is also transferred from input to output via diode D1. When Q2 is ON and Q1 is OFF, D2 provides the input-to-output energy transfer path. When both primary switches are OFF, the energy stored in the center leg is routed to the output via diode D3. The ideal voltage transfer function for this push-pull version can be determined to be:

$$\frac{V_o}{V_s} = \left[\frac{N_L}{N_P}\right]\left[\frac{D}{1 - D\left[1 - (N_L/N_s)\right]}\right] \qquad (12.48)$$

assuming that the converter is operating in a continuous output current mode. If N_L is made equal to N_s, Eq. (12.48) reduces to:

$$\frac{V_o}{V_s} = \frac{N_s}{N_P} D \qquad (12.49)$$

which is the transfer function of a quasi-squarewave buck converter, such as the circuit shown in Fig. 5.13A of Chapter 5.

In the second push-pull variation shown in Fig. 12.37B, secondary winding dot polarities are such that only magnetization energy would be stored in the center leg of its magnetic when either Q1 or Q2 is ON. Therefore, an additional winding tied to the primary source of power (N_{L1}) has been placed on the center leg to provide another source of energy storage during switch conduction

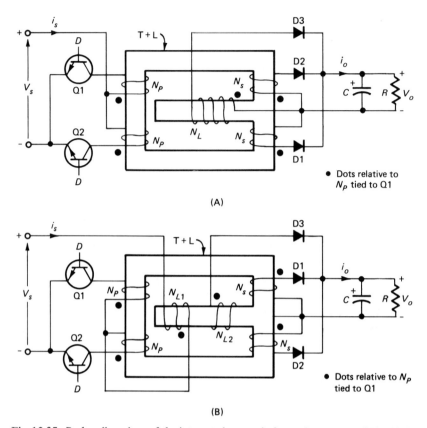

Fig. 12.37. Push-pull versions of the integrated magnetic forward converter of Fig. 12.25.

intervals. The voltage transfer function for this converter variation is then determined to be:

$$\frac{V_o}{V_s} = \frac{1}{(N_P/N_s) + (D'/D)(N_{L1}/N_{L2})} \qquad (12.50)$$

for the continuous mode of operation. If we choose to make:

$$\frac{N_P}{N_s} = \frac{N_{L1}}{N_{L2}} \qquad (12.51)$$

then the transfer function set by Eq. (12.50) becomes:

$$\frac{V_o}{V_s} = \frac{N_s}{N_P} D. \qquad (12.52)$$

This is the same transfer function that one would expect from the buck-derived Weinberg converter shown in Fig. 5.9A in Chapter 5! Thus, we can conclude that the push-pull SPC variation illustrated in Fig. 12.37B is simply an integrated magnetic version of the Weinberg converter circuit.

The various current waveforms and their timing relationships that one would expect in *either* of these two converters is illustrated in Fig. 12.38. Note that, during the times when both switches are OFF, the magnetization energy stored in the outer legs of the magnetic structure causes either D1 or D2 to conduct, thus releasing this energy to the output. The output current into the converter load and capacitance thus will be the sum of the current produced by this energy (i_o) and that flowing in D3 during these times. As shown, the total output current will normally have a slight "step" because of this summation when the two switches are turning ON or OFF. Because this energy is usually small in comparison to that stored in the center leg of the magnetic, the magnitude of this "step" will be minute and can be neglected in most instances.

Fig. 12.38B illustrates the ideal voltage waveforms at various nodes within the converters of Fig. 12.37. The time sequence of these waveforms corresponds to that of the component currents shown in Fig. 12.38A. Note that the maximum OFF voltage appearing across either of the converter's primary switches is a function of primary-to-secondary turns ratio, as well as both input and output volt-

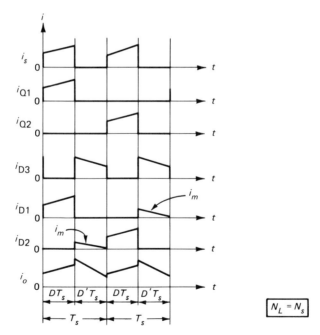

Fig. 12.38. Typical current waveforms within the converters of Fig. 12.37.

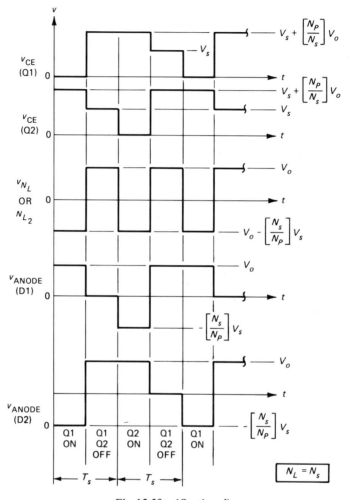

Fig. 12.38. (*Continued*)

age value. In contrast, comparable discrete-magnetic circuit counterparts, such as that shown in Fig. 5.13A in Chapter 5, will have an ideal OFF voltage stress on each primary switch that is *twice* that of the source voltage, V_s. Therefore, by judicious choice of turns ratio of transformer windings, the integrated converters of Fig. 12.37A can be purposely designed to impose *lower* OFF voltage stresses on their switches.

From a practical design standpoint, the converter version in Fig. 12.37B is preferred over that in Fig. 12.37A. Note that in version A, corresponding primary and secondary windings are on opposite sides of the core structure and therefore cannot be wound tightly together to minimize leakage inductances. By contrast, in version B just the opposite is true. Therefore, voltage transients produced by stored energy in leakage inductances during switching intervals will be much smaller in version B than in version A, thus reducing the electrical stress on converter primary switches and output diodes.

Since the converter magnetic of Fig. 12.37B contains an inductive winding in series with the outer leg primary winding, instantaneous current increase in either of the two primary switches due to conduction overlap is automatically prevented. However, this is not the case for its companion converter shown in Fig. 12.37A. Thus, the circuit of Fig. 12.37A can be considered a *voltage-fed* integrated magnetic converter, while the other version shown in Fig. 12.37B is a *current-fed* integrated magnetic converter. In fact, an integrated magnetic version of any converter circuit *must* retain *all* electrical properties of its discrete magnetic circuit counterpart to be truly contemporary in voltage and current transfer characteristics.

Using the magnetic modeling methods described earlier in this chapter, equivalent circuits for analyzing the electrical characteristics of the converters of Fig. 12.37 can easily be found. These equivalent networks are shown in Figs. 12.39 and 12.40, respectively. Note the similarity of the models of the magnetics to that shown earlier in Fig. 12.28 for the integrated forward converter. This similarity is to be expected, however, as the main inductive parts of the models should be the same; however, the basic model is changed to reflect the presence of additional windings on various legs of the magnetic cores in each instance.

Adding the provisions for two or more outputs from the two converters of Fig. 12.37 is a matter of adding more center leg and secondary windings, as is done in Fig. 12.41 for version B. This multi-output converter is an integrated

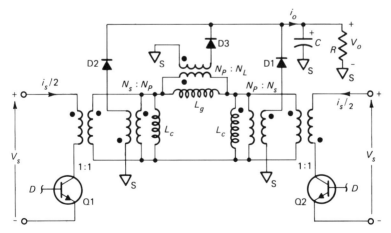

Fig. 12.39. Equivalent electric circuit model for the SPC of Fig. 12.37A.

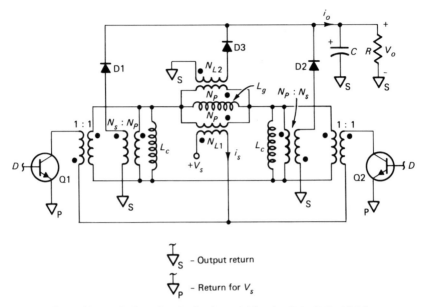

Fig. 12.40. Equivalent electric circuit model for the SPC of Fig. 12.37B.

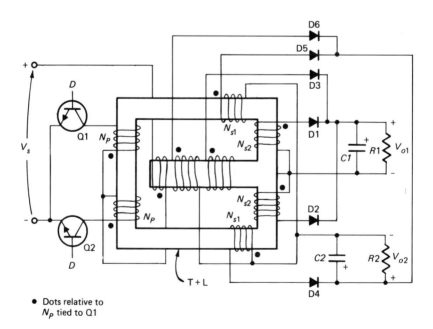

Fig. 12.41. A two-output version of the push-pull SPC shown in Fig. 12.37B.

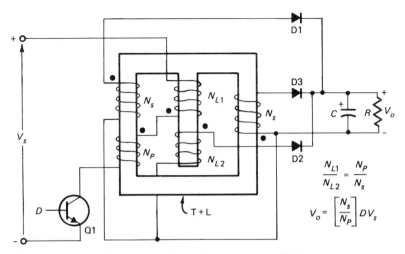

Fig. 12.42. An integrated magnetic version of the single-ended Weinberg converter shown in Fig. 5.25 of Chapter 5.

magnetic SPC whose discrete two-magnetic counterpart is the converter shown in Fig. 5.9B of Chapter 5.

With the extension of the integrated magnetic concept from single-switch to two-switch quasi-squarewave SPCs, it follows that we can apply it to other buck and boost-derived converters that have even more switches and diodes. Figs. 12.42 through 12.46 show five examples of more complicated possibilities, using

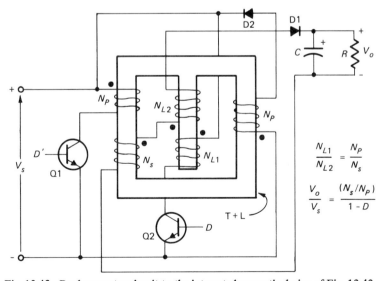

Fig. 12.43. Dual converter circuit to the integrated magnetic design of Fig. 12.42.

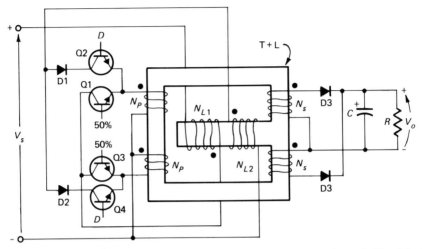

Fig. 12.44 An integrated magnetic version of the Severns converter shown in Fig. 5.8 of Chapter 5.

converter designs taken from Chapters 5, 8, and 9. Note that, in each of these five cases, even though the surrounding electronic circuit topology is more involved, the integrated magnetic structure is *still* a three-legged arrangement, with an air-gapped center leg performing the required inductive storage functions.

At this point in our work, it should be very evident that an integrated magnetic approach can be conceived for *any* buck or boost-derived SPC circuit. Although we have not specified a rigorous synthesis procedure here, evolution steps are

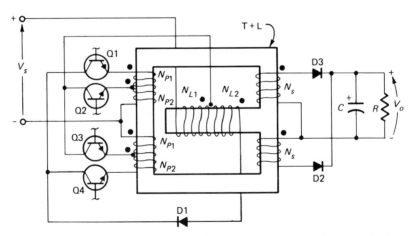

Fig. 12.45. An integrated magnetic version of the Hughes converter shown in Fig. 8.19 of Chapter 8.

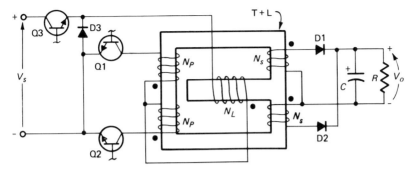

Fig. 12.46. An integrated magnetic version of the multi-mode converter shown in Fig. 9.14 of Chapter 9.

simply a matter of defining the flux relationships that exist in each magnetic component of a converter from a knowledge of the voltage and current relationships for each converter conduction state. Once these relationships are established, a magnetic structure that reproduces the flux additions or subtractions defined by the state equations is then constructed and appropriate transformer and/or inductive windings located to provide the required mmfs to produce the fluxes needed.

A more intuitive approach to synthesis is to first assume that the final magnetic structure will be three-legged, with an air-gapped center leg reserved for all inductive converter functions. All transformer windings are then placed on the outer portions of the structure, with primaries located on one outer side and secondaries on the opposite outer side. With a knowledge of the topology of the converter to be magnetically integrated, begin to interconnect the inductive and transformer-related windings until the desired magnetic relationships are satisfied. This step may require a number of iterations, but usually not many. As an aid in this interconnect procedure, assume that the inductive center leg is "isolated" from the outer legs, and that the outer legs are reserved only for transformer functions.

12.11. SMALL-SIGNAL MODELING CONSIDERATIONS

With the ability to easily develop an equivalent electrical model for any magnetic circuit regardless of its complexity, we can then apply the small-signal modeling methods of Chapter 10 to examine dynamic responses of importance for converter stability and control. The equivalent electrical model is simply substituted for the magnetic system within the converter being examined, and small-signal modeling and analysis methods applied to the resulting circuit.

For example, recall that we were able to develop equivalent circuit models for the integrated magnetic forward converter of Fig. 12.25. These models are

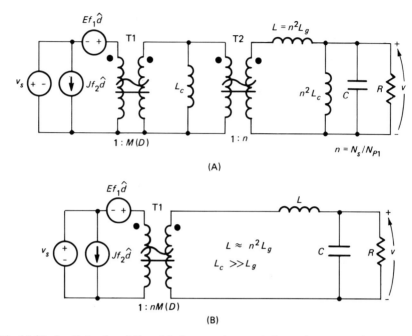

Fig. 12.47.   Small-signal modeling of the integrated magnetic forward converter of Fig. 12.25.

illustrated in Figs. 12.30A and B.   Since we were able to deduce that this particular converter was of the buck family, then the small-signal canonical model for a transformer-isolated buck SPC should apply, such as the one shown in Fig. 10.5 of Chapter 10.

Placing the various components of the electrical model of Fig. 12.30 into the proper canonical form results in the small-signal model of Fig. 12.47A.   Note that we have retained all inductive elements in this instance.   If we assume that the value of L_c is much greater than that of the air gap inductance, L_g, then a simplified canonical model can be constructed, as shown in Fig. 12.47B, which corresponds to the SPC of Fig. 12.30B.

Once again, the utility of the canonical model for SPCs becomes very evident. Addition to this model of various inductances and/or transformer elements that may be present in an integrated magnetic SPC is simple and straightforward, and their effects on small-signal responses readily found by inspection of their circuit position within the canonical model.

12.12. RIPPLE CURRENT CONTROL FOR BUCK AND BOOST CONVERTERS WITH INTEGRATED MAGNETICS

As was demonstrated in Section 12.4, auxiliary windings can be added to the inductors of basic buck and boost converters to facilitate control and reduction of input or output ripple currents. In a similar manner, windings can also be added to the center legs of the integrated magnetic versions of these converters to reduce input or output ripple current magnitudes.

For example, if a second winding is added to the gapped center leg of the magnetic used in the forward converter of Fig. 12.32, along with a DC isolation capacitor and a small external inductor for ripple trimming, then we can utilize the transformer equations of Section 12.3 to select the number of turns of this second winding so as to adjust the SPC's load current ripple to essentially zero!

The resulting integrated magnetic forward converter is shown in Fig. 12.48. The external inductor (L_{ext}) is defined to first order by the equation accompanying this figure, assuming that $N_{L1} = N_s$, and that the gap inductance (L_g) is much smaller than that of the ungapped outer legs of the magnetic structure. Both center leg windings are wound tightly together to maximize magnetic coupling between them and to minimize parasitic leakage inductances. The DC blocking capacitor, C2, is usually made large to minimize voltage ripple across it.

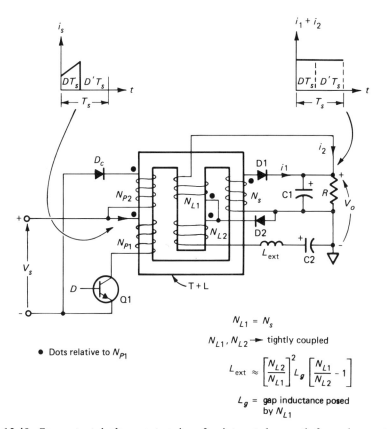

$$N_{L1} = N_s$$

$$N_{L1}, N_{L2} \rightarrow \text{tightly coupled}$$

$$L_{ext} \approx \left[\frac{N_{L2}}{N_{L1}}\right]^2 L_g \left[\frac{N_{L1}}{N_{L2}} - 1\right]$$

$$L_g = \text{gap inductance posed by } N_{L1}$$

Fig. 12.48. Zero output ripple current version of an integrated magnetic forward converter.

With it's position as shown in Fig. 12.48, this blocking capacitor will see an average DC voltage equal to that of the output potential, V_o.

The addition of this second winding does not compromise the buck input-to-output voltage gain of the forward converter, and it remains as defined earlier by Eq. (12.33). In practice, the ideal nature of the zero-slope output ripple current will not be achieved and will have some slight slope during intervals when the primary switch element (Q1 in Fig. 12.48) is ON. This slope will be proportional to the values of the inductances of the outer legs of the magnetic structure which are not controlled by the ripple-reduction center leg winding. If these induc-

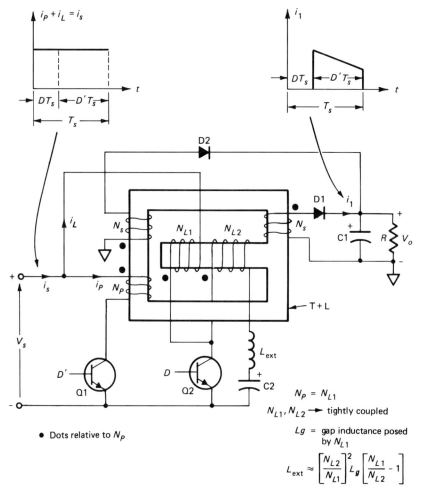

Fig. 12.49. Zero input ripple current version of an integrated magnetic boost converter.

tances are much larger than that of the gapped center leg (say 10 times), then this current deviation due to magnetization energies will be very small and can be neglected for all intensive purposes.

Fig. 12.49 shows the integrated magnetic boost converter of Fig. 12.33 altered to include a control winding to adjust the input ripple current of this SPC to zero. Once again, we can use the transformer equations of Section 12.3 to select control winding turns and the value of the external trimming inductor, as evidenced by the equation accompanying this figure. The input-to-output boost voltage gain of this converter does not change from that defined by Eq. (12.34).

In a similar fashion, we can add ripple control windings to any of the more complex integrated magnetic converters of this chapter. This is to be expected, however, since they are descendants of families parented by either the basic buck or boost converter (see Chapter 11).

12.13. CAUTIONS IN DESIGNING INTEGRATED MAGNETIC CONVERTERS

Blending of inductors and transformers of SPCs into single magnetic systems can be very advantageous, often resulting in converter designs of lower cost, weight, and size than their discrete magnetic counterparts. Conversion performance can also be improved and component stresses reduced, provided the integration process is well thought out and executed properly.

For the most part, integrating magnetic elements of a converter will not add to the many other design aspects and difficulties that an engineer must contend with during development phases. It does place additional burden on the designer to properly specify the performance of the magnetic for production use, and on its manufacturer to insure a consistent magnetic product that may be much more complex than a simple transformer or inductor. Both of these burdens add up to additional recurring and nonrecurring costs associated with an SPC. Also, changing an integrated magnetic design once it is in production can be much more costly than a discrete version because of the intimate relationships that must exist between its various inductances and/or transformer elements.

It is obvious from our modeling work of this chapter that the integration process of SPC magnetics introduces additional inductive elements in their converter circuits. A good example is the magnetizing inductances of the outer legs of the magnetic assembly (e.g., L_c in Fig. 12.39). However, by making these inductances much larger than that of the gap, their contributions to overall SPC circuit performance will be small. Nevertheless, one should always remember that they are real and finite in value, and that the energies stored in them by SPC operations must be released and dissipated in a manner so as not to harm surrounding semiconductor switches and diodes.

There are some subtle electrical problems to watch out for when coupling inductors together on a common core. In the case of a multi-output buck-derived converter with coupled output inductors (Figs. 8.4, 12.15, 12.23), imbalances in the required turns ratio relationships between inductor and transformer windings can produce circulating currents in output filter networks, resulting in high output voltage ripple. Extreme imbalance can produce high power loss in output filter capacitors due to ESR voltage drops which, in turn, could result in possible damage and even destruction of the capacitors.

Always remember when designing a coupled-inductor magnetic for a converter that the DC mmf seen by its material will be the *sum* total of winding mmfs. Therefore, it is wise to always verify that the core material of the magnetic will not be operating near saturation under maximum DC bias conditions. As we mentioned earlier in Section 12.1, operation in a nonlinear portion of the material's *B-H* loop will also result in an inaccurate electrical model that might be derived for the magnetic circuit. Also, nonlinear operation does not necessarily imply that the user circuit of the magnetic will automatically malfunction. Frequently, however, higher-than-normal inductor currents produced by core operation near saturation can cause stress or permanent damage to other converter circuit components.

When coupling the input and output inductors of a Ćuk converter to reduce current ripple (Fig. 12.7), a highly undesirable output voltage *polarity reversal* can occur when input power is first applied (Ref. 13). This transient phenomenon is illustrated in Fig. 12.50A for a transformer-isolated version of this SPC. In converter applications where short term reversals of voltage polarity could produce damage (e.g., polarized filter capacitors and monolithic integrated circuits), this phenomenon must be circumvented or reduced to acceptable magnitudes.

As shown, inrush current through the primary winding of the inductor (L1) due to first turn-on of switch Q1 or by charging of the primary energy capacitance, C_P, when V_s is first applied, will produce a secondary transient current. The direction of this current is *opposite* to that of normal current flow when the converter is operating in a steady-state fashion, and results in a *reversal* of output voltage during the period of the current transient.

This transient condition will persist until the potential voltage levels of C1 and C2 reach values to counteract that impressed across L2 by transformer action from L1. Thus, the reversal time period will be determined by the time constants and magnitudes of power stage impedances and output load values.

Figs. 12.50B and C show two viable ways of protecting loads against this transient output reversal. In part B, a blocking diode (D_B) is placed in line with L2 to insure a proper unidirectional current path for steady-state output current, but prevents reversal current flow at converter turn-on. However, its presence does lower conversion efficiency because full load current must pass through it during normal operation. In part C, a clamp diode (D_c) is added across the out-

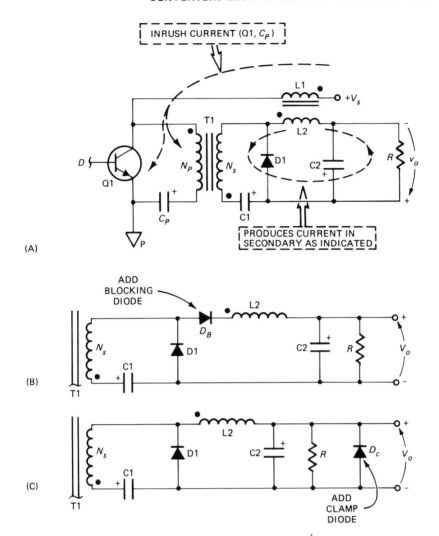

Fig. 12.50. Output voltage polarity reversal in a coupled-inductor Ćuk converter and possible methods for prevention and load protection.

put terminals of the SPC to limit the transient reversal voltage to less than a volt. During normal operation, this diode will be reverse-biased and therefore will dissipate no power. Use of a Schottky part is recommended for this diode to keep the reversal clamp voltage to an absolute minimum.

Coupling of inductors in various parts of a converter can complicate its small-signal model, often introducing additional poles or zeros in transfer functions that can produce reduced gain or even instability in a closed-loop SPC unless accounted for. However, as we saw in Chapter 10, the canonical model can be easily modified to include unregulated portions of a multi-output converter. The designer is therefore cautioned to always reflect any parts of a converter that is tied by magnetic means to a controlled loop when constructing a small-signal model for stability studies.

Finally, a few words about the design of the integrated magnetic buck and boost converters is in order before completing our exploration of integrated magnetics for SPCs. The amount of energy to be stored in the center legs of the designs that we have shown in this chapter will, of course, be dependent on the total output power needs of their converters. This center leg also will have significant DC bias because the unidirectional nature of its flux. Since the dimensions of a ferromagnetic material can be directly related to energy storage capability, one should attempt to estimate the size of the center leg of the magnetic from a knowledge of the energy needed and the rate at which it must be supplied. This estimate can then be used to select a magnetic material, core size/shape, turns and air gap lengths that will produce a design that meets or exceeds the energy storage requirement. For those interested in the details of magnetic field energy and its relation to core volume and material characteristics, Ref. 9 listed in the bibliography section of this book is again highly recommended for supplementary reading.

Bibliography

TEXT REFERENCES

1. Middlebrook, R. D. and Ćuk, S., "A General Unified Approach to Modelling Switching-Converter Power Stages," *IEEE Power Electronics Specialists Conference Record*, June 1976, pp. 18-34.
2. Middlebrook, R. D. and Ćuk, S., "A General Unified Approach to Modelling DC-to-DC Converters in Discontinuous Conduction," *IEEE Power Electronics Specialists Conference Record*, June 1977, pp. 36-57.
3. Lin, P. M. and Chua, L. O., "Topological Generation and Analysis of Voltage Multiplier Circuits," *IEEE Transactions on Circuits and Systems*, Volume CAS-24, Number 10, October 1977, pp. 517-530.
4. Ninomiya, T., Harada, K., and Nakahara, M., "On the Maximum Regulation Range in Boost and Buck Converters," *IEEE Power Electronics Specialists Conference Record*, June 1981, pp. 146-153.
5. Hnatek, E., *Design of Solid State Power Supplies*, Van Nostrand Reinhold Co., New York, 1971, pp. 32-33.
6. Ćuk, Slobodan, "General Topological Properties of Switching Structures," *Power Electronics Specialists Conference Record*, June 1979, pp. 109-130.
7. Middlebrook, R. D., "Input Filter Considerations in Design and Application of Switching Regulators," *IEEE Industrial Applications Society Annual Meeting Record*, October 1976, pp. 366-382.
8. Middlebrook, R. D., "Design Techniques for Preventing Input Filter Oscillations in Switched-Mode Regulators," *Proceedings of POWERCON 5*, May 1978, pp. A3-1 to A3-16.
9. Watson, J. K., *Applications of Magnetism*, John Wiley & Sons, New York, 1980, pp. 133-140, 273-278.
10. Rensink, L., *Switching Regulator Configurations and Circuit Realizations*, Ph. D. Thesis, California Institute of Technology, 1980, pp. 56-66.
11. Ćuk, Slobodan, "Analysis of Integrated Magnetics to Eliminate Current Ripple In Switching Converters," *PCI Conference Proceedings*, April, 1983, pp. 361-386.
12. J. Cielo and H. Hoffman, "Combined Transformer and Indicator Device," U.S. Patent 3,553,620 (January 5, 1971).
13. Bloom, G. E. and Eris, A., "Practical Design Considerations of A Multi-Output Ćuk Converter," *IEEE Power Electronics Specialists Conference Record*, June 1979, pp. 133-146.

FIGURE REFERENCES

The references in this section are organized by the figure number. For each figure where the authors are aware of a prior source or reference, that source is listed. Circuits, etc., for which no references are given are, to the best of the authors' knowledge, either original creations or are designs whose origins are obscure.

4.11.   Royer, G. H., "A Switching Transistor DC-to-AC Converter Having An Output Frequency Proportional to the DC Input Voltage," *AIEE Proceedings*, July 1955.

4.12.   Jensen, J., "An Improved Square Wave Oscillator Circuit," *IRE Transactions On Circuit Theory*, September 1957.

5.3.   Cronin, D., et al., "Multiple High Voltage Output DC-to-DC Power Converter," U.S. Patent 4,034,280 (July 5, 1977).

Farber, D., et al., "Power Processing System for a 200W Communication Satellite Transmitter," International Conference on Satellite Communication Systems Technology, April 7–10, 1975.

Farber, B., et al., "A High Power TWT Power Processing System," *IEEE Power Electronics Specialists Conference Record*, June 1974.

5.5C.   Higuchi, H. and Trubell, L., "Regulating Electric Power Circuit Arrangement," U.S. Patent 4,025,863 (May 24, 1977).

5.6.   Disclosed by D. Cronin to R. Severns, TRW Inc., January 1978. Actual development work done at an earlier unknown date.

5.7.   Hunter, I. and Fitzgerald, W., "Regulated Direct Current Supply Circuit with Energy Return Path," U.S. Patent 3,432,737 (March 11, 1969).

De la Lastra, A., "Power Converter for Converting an Unregulated DC Input into a Regulated Output Voltage," U.S. Patent 3,388, 311 (June 11, 1968).

5.8.   Severns, R., "A New Current-Fed Converter Topology," *IEEE Power Electronics Specialists Conference Record*, June 1979.

5.9.   Weinberg, A. H., "A Boost Regulator With a New Energy Transfer Principle," Proceedings of the Spacecraft Power Conversion Electronics Seminar, September 1974 (ESRO Publication SP-103).

Biess, J., et al., "Power Processing Module for Military Digital Equipment Power Sub-System," *IEEE Power Electronics Specialists Conference Record*, June 1977.

Cronin, D. and Biess, J., "Series Induction/Parallel Inverter Power Stage and Power Staging Method For DC-to-DC Power Converter," U.S. Patent 4,176,392 (November 27, 1979).

Higgins, E. R., "Direct Current To Direct Current Chopper Inverter," U.S. Patent 3,846,691 (November 5, 1974).

5.10.   Bloom, G. E., "Optimum Topology High Voltage DC-to-DC Converter," U.S. Patent 4,318,166 (March 2, 1982).

5.16B. Calkin, E. and Hamilton, B., "A Conceptually New Approach for Regulated DC-to-DC Converters Employing Transistor Switches and Pulse Width Control," IEEE Industry Applications Society Conference Record, 1972.

Millman, J., "Designing High Performance Power Converters for an Airborne Radar TWT Transmitter," Solid-State Power Conversion Magazine, November/December 1978.

Calkin, E., et al., "Regulated DC-to-DC Converter with Regulated Current Source Driving a Non-Regulated Inverter," U.S. Patent 3,737,755 (June 5, 1973).

5.17   Severns, R., "Switchmode Converter Topologies–Make Them Work For You!," Intersil, Inc., Application Note A035, 1980 (Fig. 33).

5.20B. VanVelthooven, K. and Koppe, H., "Low Cost Forward Converters Ease Switching Supply Design," *Electronics Magazine*, February 2, 1978.
Hayes, S., "A Design Technique For Optimizing The Power Device Utilization in Feed-Forward Converters," *Proceedings of POWERCON 8*, March 1981.

5.21B. See references for Figure 5.20B.

5.22A. Lillienstein, M. and Miller, R., "The Biased Transformer DC-to-DC Converter," *IEEE Power Electronics Specialists Conference Record*, June 1976.

5.22B. Kamata, Y. and Katsou, K., "DC-to-DC Converter," U.S. Patent 3,935,526 (January 27, 1976).

5.25.   See references for Fig. 5.17.

6.3.   Severns, R., "Techniques for Designing New Types of Switching Regulators," *European Power Conversion Conference Record*, September 1979.

6.4B.   Clark, P. W., "Converter Regulation by Controlled Conduction Overlap," U.S. Patent 3,938,024 (February 10, 1976).

6.5.   See references for Fig. 5.17.

6.6A.   Porter W. T., "Frequency Modulated Switching Regulator," *IEEE Power Electronics Specialists Conference Record*, June 1979.

6.6B.   Severns, R., "Switchmode and Resonant Power Converter Circuits," International Rectifier Applications Note, 1981.

6.7.   See references for Fig. 5.17.

6.8.   See references for Fig. 6.6B.

6.10A. See references for Fig. 6.6B.

6.13.   See references for Fig. 6.6B.
Bloom, G. E. and Severns, R., "Unusual DC-DC Power Conversion Systems," *1980 MIDCON Professional Program Record*, 1980.

7.2.   Matsuo, H. and Harada, K., "The Cascade Connection of Switching Regulators," *IEEE Transactions on Industry Applications*, Vol. 1A-12, Number 2, March/April 1976.

7.6.   Ćuk, Slobodan, "Modelling, Analysis and Design of Switching Regulators," Ph.D. Thesis, California Institute of Technology, November, 1976. Also NASA Report CR-13514.
Middlebrook, R. D. and Ćuk, S., "A New Optimum Topology Switching DC-to-DC Converter," *IEEE Power Electronics Specialists Conference Record*, June 1977.
Ćuk, S. and Middlebrook, R. D., "DC-to-DC Switching Converter," U.S. Patent 4,184,197 (January 15, 1980). Foreign patents also known to be pending.
Ćuk, Slobodan, "Discontinuous Inductor Current Mode in the Optimum Topology Switching Converter," *IEEE Power Electronics Specialists Conference Record*, June 1978.

7.8.   Middlebrook, R. D. and Ćuk, S., "Isolation and Multiple Output Extensions of a New Optimum Topology Switching DC-to-DC Converter," *IEEE Power Electronics Specialists Conference Record*, June 1978.
Bloom, G. E. and Eris, A., "Practical Design Considerations of a Multi-Output Ćuk Converter," *IEEE Power Electronics Specialists Conference Record*, June 1979.

7.9A.   Ćuk, S. and Middlebrook, R. D., "Coupled-Inductor and Other Extensions of a New Optimum Topology Switching DC-to-DC Converter," *IEEE Industry Applications Society Meeting Record,* October 1977.
Erickson, W. and Ćuk, S., "A Conceptually New High Frequency Switched-Mode Amplifier Technique Eliminates Current Ripple," *Proceedings of POWERCON 5*, May 1978.

       Ćuk, Slobodan, "Switching DC-to-DC Converter with Zero Input or Output Current Ripple," *IEEE Industry Applications Society Annual Meeting Record*, October 1978.

7.9C.   Bloom, G. E. and Eris, A., "DC-to-DC Converter," U.S. Patent 4,262,328 (April 14, 1981).

       Bloom, G. E. and Eris, A., "DC-to-DC Converter," U.S.Patent 4,355,352 (October 19, 1982).

       Bloom, G. E. and Eris, A., "Practical Design Considerations for a Multi-Output Ćuk Converter," *IEEE Power Electronics Specialists Conference Record*, June 1979.

       Ćuk, S. and Middlebrook, R. D., "DC-to-DC Converter Having Reduced Ripple Without Need For Adjustments," U.S. Patent 4,274,133 (June 16, 1981).

       Rensink, L., *Switching Regulator Configurations and Circuit Realizations*, Ph.D. Thesis, California Institute of Technology, June 1980.

7.10.   Landsman, E., "A Unifying Derivation of Switching Regulator Topologies," *IEEE Power Electronics Specialists Conference Record*, June 1979.

       Ćuk, S. and Middlebrook, R. D., "DC-to-DC Converter Having Reduced Ripple Without Need For Adjustments," U.S. Patent 4,274,133 (June 16, 1981).

7.11.   See references for Fig. 7.9.

7.12B. Massey, R. P. and Snyder, E. C., "High Voltage Single-Ended DC-DC Converter," *IEEE Power Electronics Specialists Conference Record*, June 1977.

7.13.   Ćuk, Slobodan, "General Topological Properties of Switching Structures," *IEEE Power Electronics Specialists Conference Record*, June 1979.

       Clarke, P. W., "A New Switched-Mode Power Conversion Topology Provides Inherently Stable Response," *Proceedings of POWERCON 10*, March 1983.

7.17.   Khno, M. and Kuwabara, K., "Single-Ended DC-to-DC Converter With Two Individually Controlled Outputs," *IEEE International Telecommunications Energy Conference Record*, November, 1979.

8.3B.   Frosch, R. A. and McLyman, W. T., "Elimination of Current Spikes in Buck Power Converters," U.S. Patent 4,245,288 (January 13, 1981).

       McLyman, Colonel Wm. T., *NASA Technical Brief*, Vol. IV, Number 3, Fall 1979.

       Webster, G. W., "An Improved Push-Pull Voltage Fed Converter Using a Tapped Output-Filter Inductor," *IEEE Power Electronics Specialists Conference Record*, June 1983.

8.4.   Lloyde, A., "Choking up on LC Filters," *Electronics Magazine*, August 21, 1967.

       Waehner, G. C., "Switching Power Supply Common Output Filter," U.S. Patent 3,916,286 (October 28, 1975).

       Hirschberg, W. J., "Improving Multiple Output Converter Performance With Coupled Output Inductors," *Proceedings of POWERCON 9*, July 1982.

8.5A.   Crouse, G. B., "Electrical Filter," U.S. Patent 1,920,948 (August 1, 1933).

       Feng, S., et al., "Small-Capacitance Nondissipative Ripple Filters for DC Supplies," *IEEE Transactions on Magnetics*, Vol. MAG-6, Number 1, March 1970.

8.7A.   Maresca, T. J., "Regulated DC-to-DC Converter," *IEEE Transactions on Magnetics*, March 1970.

8.18.   Disclosed to G. E. Bloom by A. Eris, Litton Guidance & Control Systems, May 6, 1981. Patents may be pending.

8.20B. Venable, H. D., "Regulated DC to DC Converter," U.S. Patent 3,925,715 (December 9, 1975).

       Hayner, R., et al., "The Venable Converter: A New Approach to Power Processing," *IEEE Power Electronics Specialists Conference Record*, June 1976.

Rostad, A., et al., "Application of the Venable Converter to a Series of Satellite TWT Power Processors," *IEEE Power Electronics Specialists Conference Record*, June 1976.

Venable, H. D., "Designing the Off-Line, High Efficiency, Venable Converter," *Proceedings of POWERCON 6*, May 1979.

9.1. Cardwell, G. and Neel, W., "Bilateral Power Conditioner," *IEEE Power Electronics Specialists Conference Record*, June 1973.

Matsuo, H. and Harada, K., "New DC-DC Converters with an Energy Storage Reactor," *IEEE Transactions on Magnetics*, Vol. MAG-13, Number 5, September 1977.

Middlebrook, R. D., et al., "A New Battery Charger/Discharger Converter," *IEEE Power Electronics Specialists Conference Record*, June 1978.

9.11. See references for Fig. 5.8.

9.12. See references for Fig. 5.8.

9.13. This circuit was developed by R. Severns while under contract to Xerox Corporation, El Segundo, California, in October 1980.

9.14. See references for Fig. 5.3.

9.16. See references for Fig. 5.9.

9.17. See references for Fig. 8.20B.

10.10. Webster, G. W., "An Improved Push-Pull Voltage Fed Converter Using a Tapped Output-Filter Inductor," *IEEE Power Electronics Specialists Conference Record*, June 1983.

12.1. Hunt, W. T., Jr. and Stein, R., *Static Electromagnetic Devices*, Allyn and Bacon, Inc., Boston, 1963, pp. 121, 122.

12.4. Watson, J. K., *Applications of Magnetism*, John Wiley & Sons, New York, 1980, pp. 277, 278.

12.5. See references for Fig. 12.4.

12.6. See reference for Fig. 12.1, page 50.

12.7. See references for Fig. 7.9A.

12.16. See references for Fig. 7.9C.

12.17. Rensink, L., *Switching Regulator Configurations and Circuit Realizations*. Ph.D. Thesis, California Institute of Technology, 1980, page 64.

12.24. Ćuk, Slobodan, "Analysis of Integrated Magnetics to Eliminate Current Ripple In Switching Converters," *PCI Conference Proceedings*, April 1983.

12.24. Bloom, G. E., et al., "Modeling and Analysis of a Milti-Output Ćuk Converter," *IEEE Power Electronics Specialists Conference Record*, June 1980.

12.25. Bloom, G. E. and Eris, A., "Practical Design Considerations of a Multi-Output Ćuk Converter," *IEEE Power Electronics Specialists Conference Record*, June 1979.

Suggested Reading and Other Information Sources

Unfortunately, for the reader interested in the pursuit of previous work done in the field of switched-mode DC-to-DC power converter design, there is no single information source available today that can be considered comprehensive in its coverage of this subject. As witness the variety of publications given in the reference section of this text, most of the useful design information is scattered in books, magazines and in the proceedings of a host of conferences and other technical symposia, many of which are not directly related to the aspects of power converter design.

However, there are information sources available that are considered essential references for anyone beginning a career in the field of power electronics, as well as for those individuals engaged in serious research on related subjects. The purpose of this section is to provide a listing of these information sources, along with other data important for location and procurement.

1. CONFERENCE RECORDS

Perhaps the best-known technical meeting concerned with the subjects of power electronics is the annual Power Electronics Specialists Conference (PESC). Currently sponsored by the Aerospace and Electronic Systems Society of the Institute of Electrical and Electronic Engineers (IEEE), this conference brings together specialists in circuits, systems, electron devices, magnetics, control theory, instrumentation and power engineering for interspeciality discussions of new ideas, research and development, applications, and the latest advances in the field of power electronics. The conference was first held in 1970, and its technical programs consist of papers which describe original and unpublished work, both experimental and analytical. Records of all conferences may be obtained at a nominal cost from the IEEE Services Center, 445 Hoes Lane, Piscataway, New Jersey, USA, 08854. IEEE catalog numbers are given below for records from 1970 to 1984:

Year	IEEE Catalog Number
1970	70C31-AES
1971	71C15-AES
1972	72CHO-652-8 AES
1973	73CHO-787-2 AES

Year	IEEE Catalog Number
1974	74CHO-863-1 AES
1975	75CHO-965-4 AES
1976	76CH-1084-3 AES
1977	77CH-1213-8 AES
1978	78CH-1337-5 AES
1979	79CH-1461-3 AES
1980	80CH-1529-7 AES
1981	81CH-1652-7 AES
1982	82CH-1762-4 AES
1983	83CH-1877-0 AES
1984	84CH-2000-8 AES

Another IEEE meeting of particular interest to individuals concerned with power semiconductor devices and their application is the International Semiconductor Power Converter Conference (ISPCC). This conference was first held in 1952 and has been repeated every five years since then. The ISPCC is sponsored by the Industry Applications Society (IAS) of the IEEE. As with the PESC records, the service office of the IEEE should be contacted on availability of past conference proceedings. At the time of publication of this text, the last ISPCC meeting was held in 1982 and its IEEE catalog record number is 82CH-1682-4 IAS.

Aside from the IEEE-sponsored PESC and ISPCC meetings, there are two other conferences related to power conversion that are frequented by power electronics engineers. Both are held annually and both are sponsored by United States-based corporations. The oldest of these two conferences is POWERCON, with its first forum held in 1974. While all POWERCON conferences have been held to date in various parts of the United States, the other corporation-sponsored conference is often conducted in cities of European countries as well as in the United States. This latter conference is called the Power Conversion International Conference (PCI). The first PCI conference was held in Munich, West Germany, in 1979.

POWERCON is sponsored by Power Concepts, Incorporated, and is a registered service mark of that corporation. Persons interested in purchasing copies of conference records should contact the POWERCON business office at Post Office Box 5226, Ventura, California USA, 93003 for information on cost and availability.

The PCI conference is sponsored by Interlec Communications, Incorporated. Questions on purchase and availability of conference records should be directed to their business office at 2909 Ocean Drive, Oxnard, California USA, 93030.

There are many other worldwide society-sponsored conferences that include power electronic subjects as a part of their forum. Examples are the International Telecommunications Energy Conference and the annual IEEE-sponsored Industry Applications Society (IAS) meeting. Most of these conferences are coordinated through the IEEE in the United States; therefore, it is recommended that the service office of IEEE be contacted to learn about such conferences and on how to obtain proceeding records. The address of the IEEE service office is given above in the discussion about the PESC proceedings.

2. TECHNICAL BOOKS

For most persons (authors of this text included) now actively engaged in the development of power converters, proper methods of design practice were learned by on-the-job industrial rather than engineering college experiences. It is widely recognized within the community of power electronics engineers today that the lack of formal college educational opportunities in power electronics has hampered progress and technical advancements in these fields. This educational dilemma is also reflected by the lack of fundamental textbooks containing subject matter related to power electronics circuit design, analysis, and synthesis.

However, over the past ten years, a limited number of technical books devoted to power electronics components, circuits and systems have appeared on the scene. The following is a list of those texts which the authors feel are useful reading materials for both the novice and the professional power electronics engineer. In addition to this list, the reader is also encouraged to obtain copies of the texts listed in the Bibliography.

1. Roddam, T., *Transistor Inverters and Converters*, D. Van Nostrand Company, Inc., Princeton, NJ, 1963.
2. Pressman, A. I., *Switching And Linear Power Supply, Power Converter Design*, Hayden Book Company, Inc., Rochelle Park, New Jersey, 1977.
3. Devan, S. B. and Straughen, A., *Power Electronics Circuits*, John Wiley & Sons, New York, 1975.
4. Stein, R., and Hunt, W. T., Jr., *Electric Power System Components*, Van Nostrand Reinhold Company, New York, 1979.
5. Wood, P., *Switching Power Converters*, Van Nostrand Reinhold Company, New York, 1981.
6. Middlebrook, R. D. and Ćuk, S., *Advances in Switched-Mode Power Conversion*, TESLAco, Inc., Pasadena, California, 1981 (multi-volume series).
7. McLyman, Colonel Wm. T., *Transformer and Inductor Design Handbook*, Marcel Dekker, Inc., New York, 1978.
8. McLyman, Colonel Wm. T., *Magnetic Core Selection for Transformers and Inductors*. Marcel Dekker, Inc., New York, 1982.
9. Bedford, B. D. and Hoft, R. G., *Principles of Inverter Circuits*, John Wiley & Sons, New York, 1964.

The multi-volume series published by TESLAco (No. 6 above) is particularly useful, as it contains the majority of the papers published by the Power Electronics Group at the California Institute of Technology (CALTECH) since 1970. Included in these volumes are many of the CALTECH papers listed earlier as text references for this book. Information on price and availability of the volumes can be obtained by contacting the TESLAco business office at 490 South Rosemead Blvd., Suite #6, Pasadena, California 91107 USA.

3. MAGAZINES, JOURNALS, AND PERIODICALS

The current plight of the power electronics engineer in his search for regular and periodic information on subjects important to his field is perhaps exemplified by the lack of periodicals, journals or magazines devoted to such subject matter. At the time of the publication of this book, only one of the magazines listed below is considered by many subscribers to be one that is directed to power electronics designers. With respect to the others, they have, at irregular intervals, published technical articles of interest to power conversion design engineers. The authors of this text suggest that readers interested in obtaining copies of such articles contact the editors of each publication and request available index listings.

Publication	Publisher/Editor Contact
1. *Powerconversion International Magazine*	Intertec Communications, Inc. 2909 Ocean Drive Oxnard, CA 93030 USA
2. *IEEE Transactions on Aerospace and Electronic Systems*	IEEE Service Office 445 Hoes Lane Piscataway, NJ 08854 USA
3. *IEEE Transactions on Magnetics*	Same contact as above.
4. *IEEE Transactions on Industry Applications*	Same contact as above.
5. *EDN Magazine*	Cahners Publishing Co. 221 Columbus Avenue Boston, MA 02116 USA
6. *Electronic Design Magazine*	Hayden Publishing Co. 50 Essex Street Rochelle Park, NJ 07662 USA
7. *Electronic Products Magazine*	Hearst Business Communications, Inc. UTP Division 645 Stewart Avenue Garden City, NY 11530 USA
8. *Wireless World*	IPC Electrical-Electronic Press Ltd., Quadrant House, The Quadrant, Sutton Surrey ENGLAND SM2 5AS
9. *Electronics Magazine*	McGraw-Hill Publishers 305 Madison Avenue New York, NY 10017 USA

4. PATENT FILES

Perhaps the most thought-provoking sources of information on power converter design are foreign and domestic patent records. At the time of the publication of this book, there were well over 4,300,000 patents on file in the United States Patent Office alone, with an average patent application submittal rate on the order of 100,000 per year.

Because of the enormous number of patents (on everything from the safety pin to Nikola Tesla's electromagnetic motor concept), patent records have special formats and codings to distinguish various invention categories. Many of these categories deal with power conversion circuit concepts and, therefore, it is not difficult (although somewhat time-consuming) for an interested researcher to locate them during a patent file search.

A complete file of all United States patents (including the first one issued in 1790 to Samuel Hopkins for an improvement in "The making of Pot ash and Pearl ash by a new Apparatus and Process") is maintained by the United States Patent and Trademark Office in Arlington, Virginia. The office is located at Crystal Plaza, 2021 Jefferson Davis Highway, in Arlington. Copies of patents can also be obtained from this office at a nominal cost. For those persons wishing to obtain a detailed guide to the standards for patent drawings, etc., the Patent and Trademark Office in Washington, D.C. 20231 can provide a copy, again at a small cost.

In addition to the central patent record office in Virginia, many major United States cities have public libraries that maintain copies and records of patents. Therefore, it is recommended that your local public library be contacted to see if they have patent files available for viewing and copying, before going directly to the U.S. Patent Office. Also, there are many patent information data banks that are accessible by computer telecommunication methods today. Many companies subscribe to such information services and your employer might be one of them.

Finally, for those readers who might enjoy a light-hearted look at the history of patents plus a guide to the preparation of a patent, *The Patent Book*, by James Gregory and Kevin Mulligan, is highly recommended. This book was first published in 1979 by A & W Publishers, Inc., 95 Madison Avenue, New York, New York 10016 USA, and its Library of Congress Catalog Card Number is 78-68388.

Index